I0503902

Copyright notice:

A note about Haeckel's work:

When I first came across a drawing by Ernst Haeckel I was astonished. I am neither a biologist nor an anatomist and the detail surpassed anything I had seen before. A little research on the man who created the drawing I was looking at (reproduced as the cover of this book) turned up the following information on Wikkipedia:

"**Ernst Heinrich Philipp August Haeckel** (February 16, 1834 – August 9, 1919) was a German biologist, naturalist, philosopher, physician, professor and artist who discovered, described and named thousands of new species, mapped a genealogical tree relating all life forms, and coined many terms in biology, including *anthropogeny, ecology, phylum, phylogeny, stem cell*, and the kingdom *Protista.* Haeckel promoted and popularized Charles Darwin's work in Germany and developed the controversial recapitulation theory ("ontogeny recapitulates phylogeny") claiming that an individual organism's biological development, or ontogeny, parallels and summarizes its species' evolutionary development, or phylogeny.

"The published artwork of Haeckel includes over 100 detailed, multi-color illustrations of animals and sea creatures (see: *Kunstformen der Natur*, "Art Forms of Nature")." Many reproductions of this set of illustrations can be found on Amazon.com. One of the features of most of the reproductions of Haeckel's works is that the images have been "cleaned up". That is, they have been extracted from the images of the original aging paper along with the German script identifying the plate (*Tafel*) number, and many have been "Photoshopped" to enhance their color or contrast. I have elected to reproduce the images here without any enhancements so that what we are seeing is likely closer to the current condition of the originals.

I fully expect to see some of these drawings on T-shirts, coffee mugs and perhaps even the hoods of sports cars. I'm not certain the good professor would object to this. He appears to have taken some liberties with some of his own drawings and was at the center of controversies in which he was charged with being a fraud. But that's another story. In these pages one can only marvel at nature and at the hand which created these images.

David Abbey, PhD
Guelph, Ontario
Canada

August, 2014

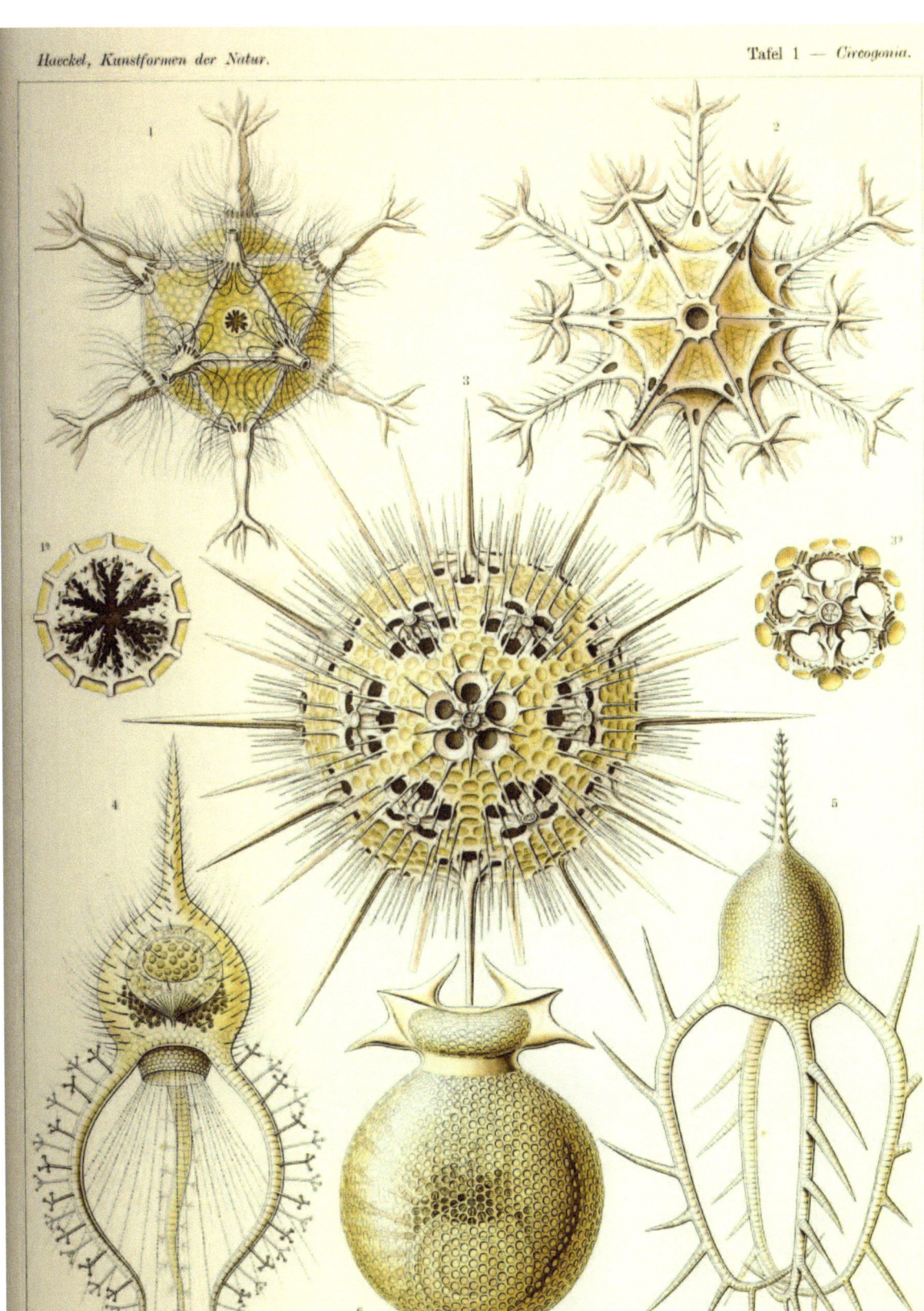

Phaeodaria. — Rohrstrahlinge.

Thalamophora. — Kammerlinge.

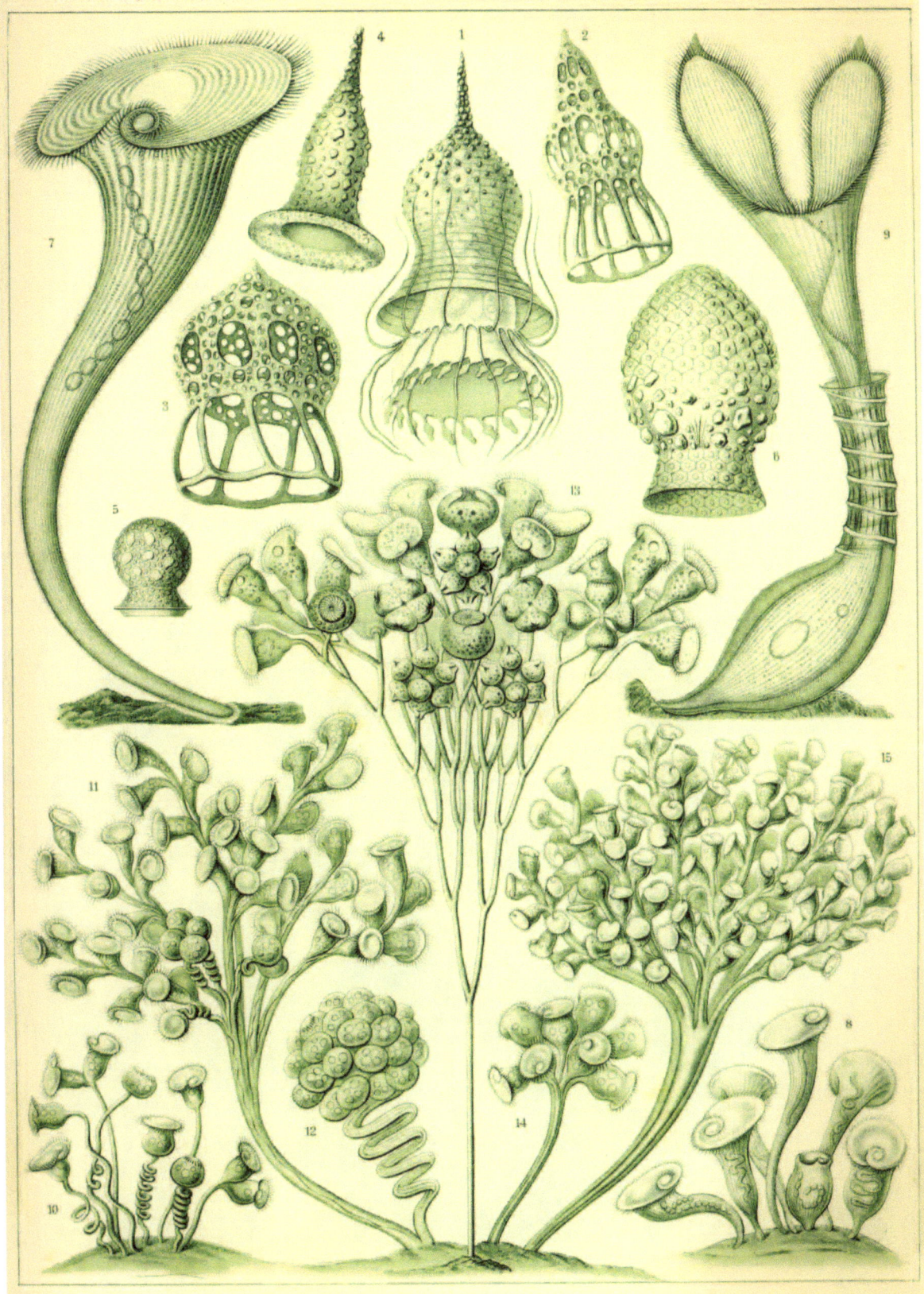

Ciliata. — Wimperlinge.

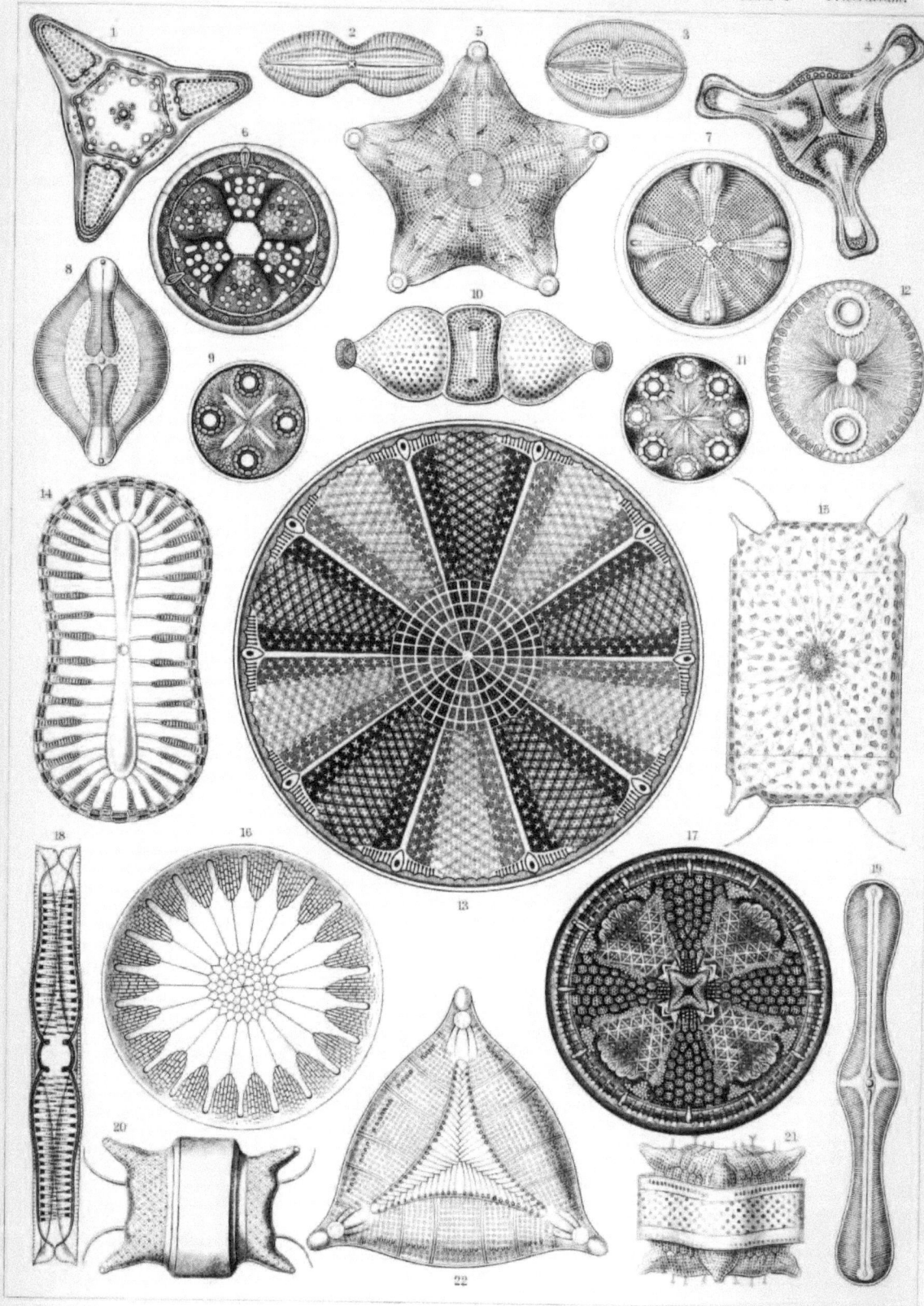

Diatomea. — Schachtellinge.

Calcispongiae. — Kalkschwämme.

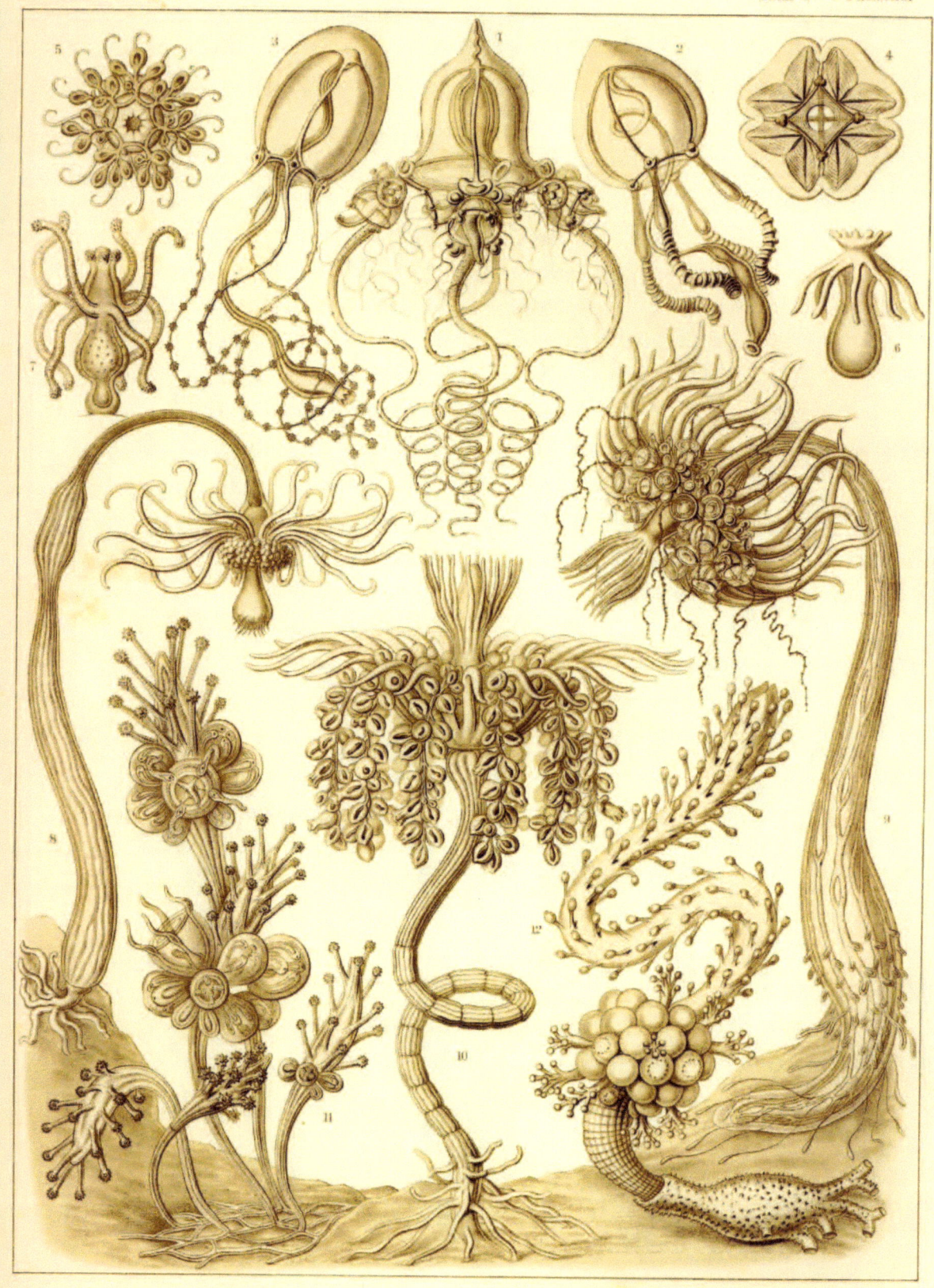

Tubulariae. — Röhrenpolypen.

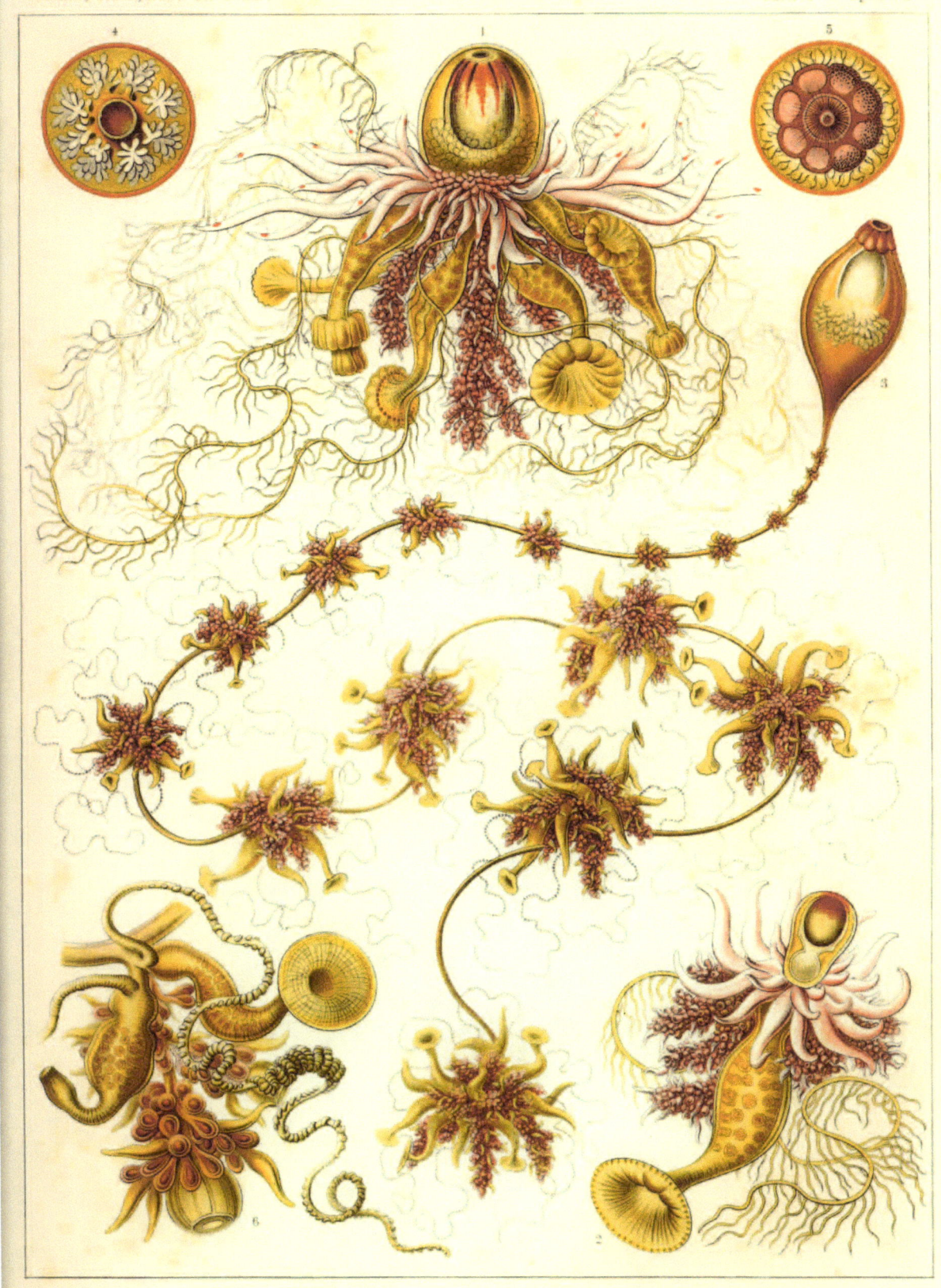

Siphonophorae. — Staatsquallen.

Discomedusae. — Scheibenquallen.

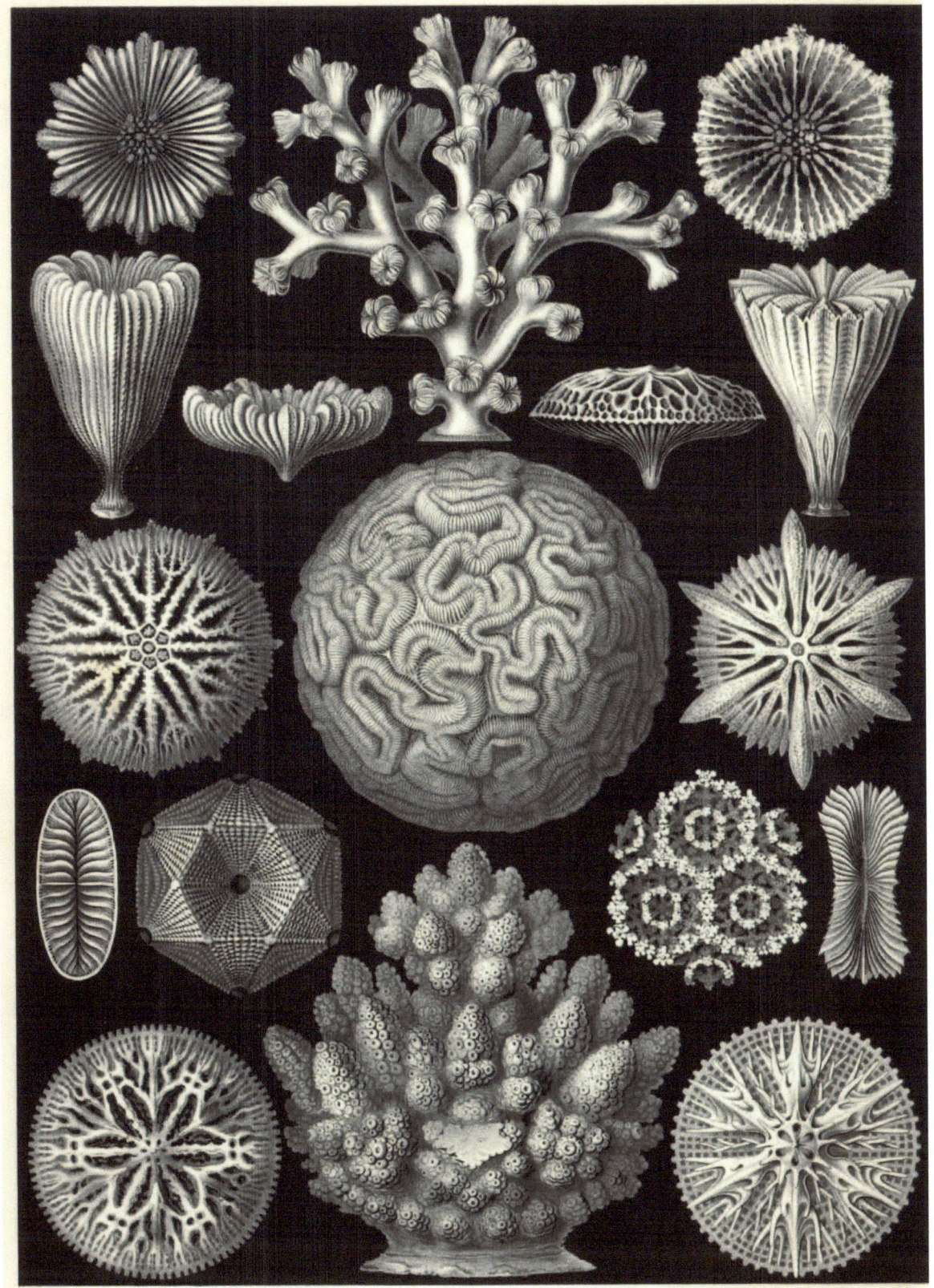

Hexacoralla. — Sechsstrahlige Sternkorallen.

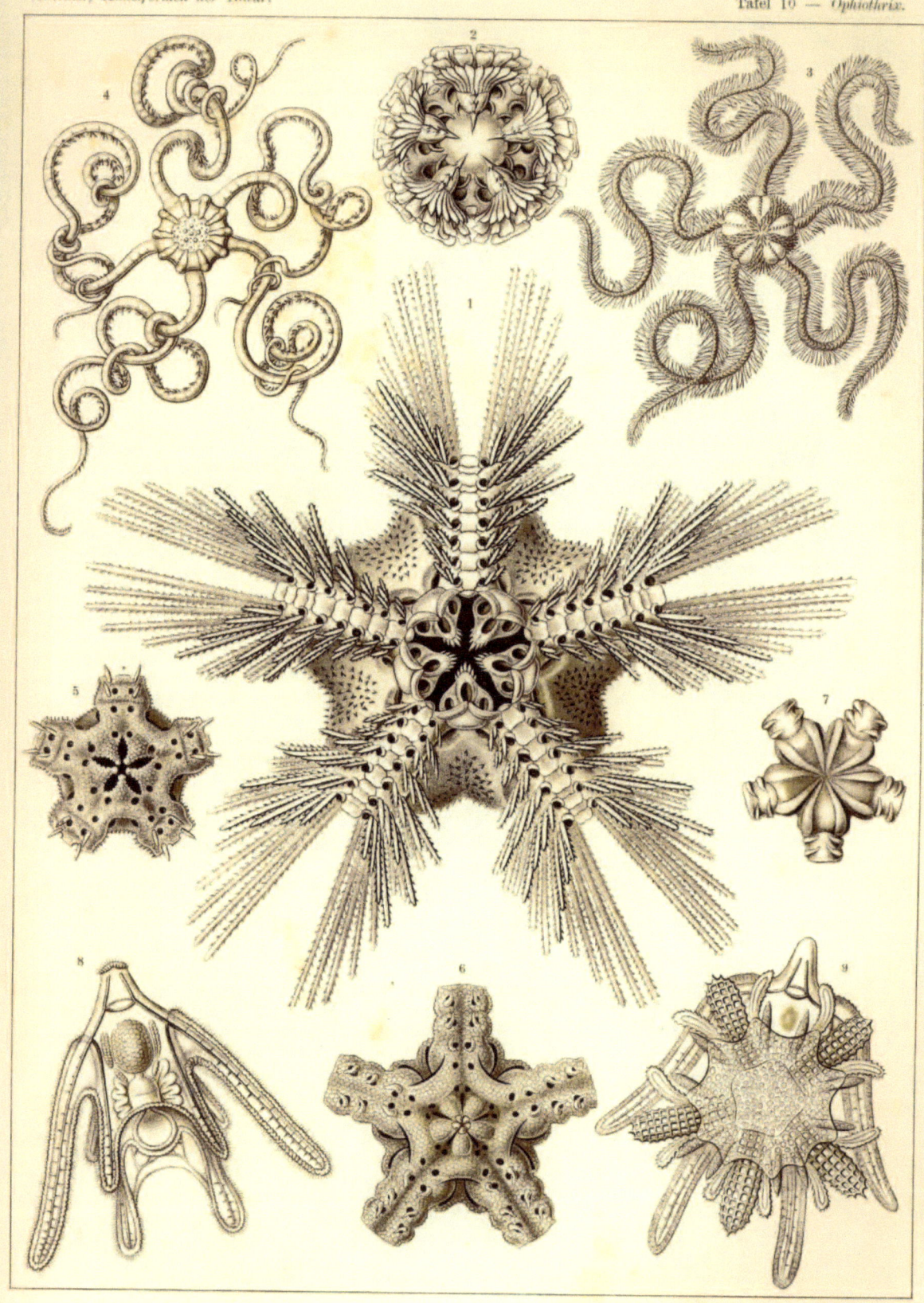

Ophiodea. — Schlangensterne.

Discoidea. — Scheiben-Strahlinge.

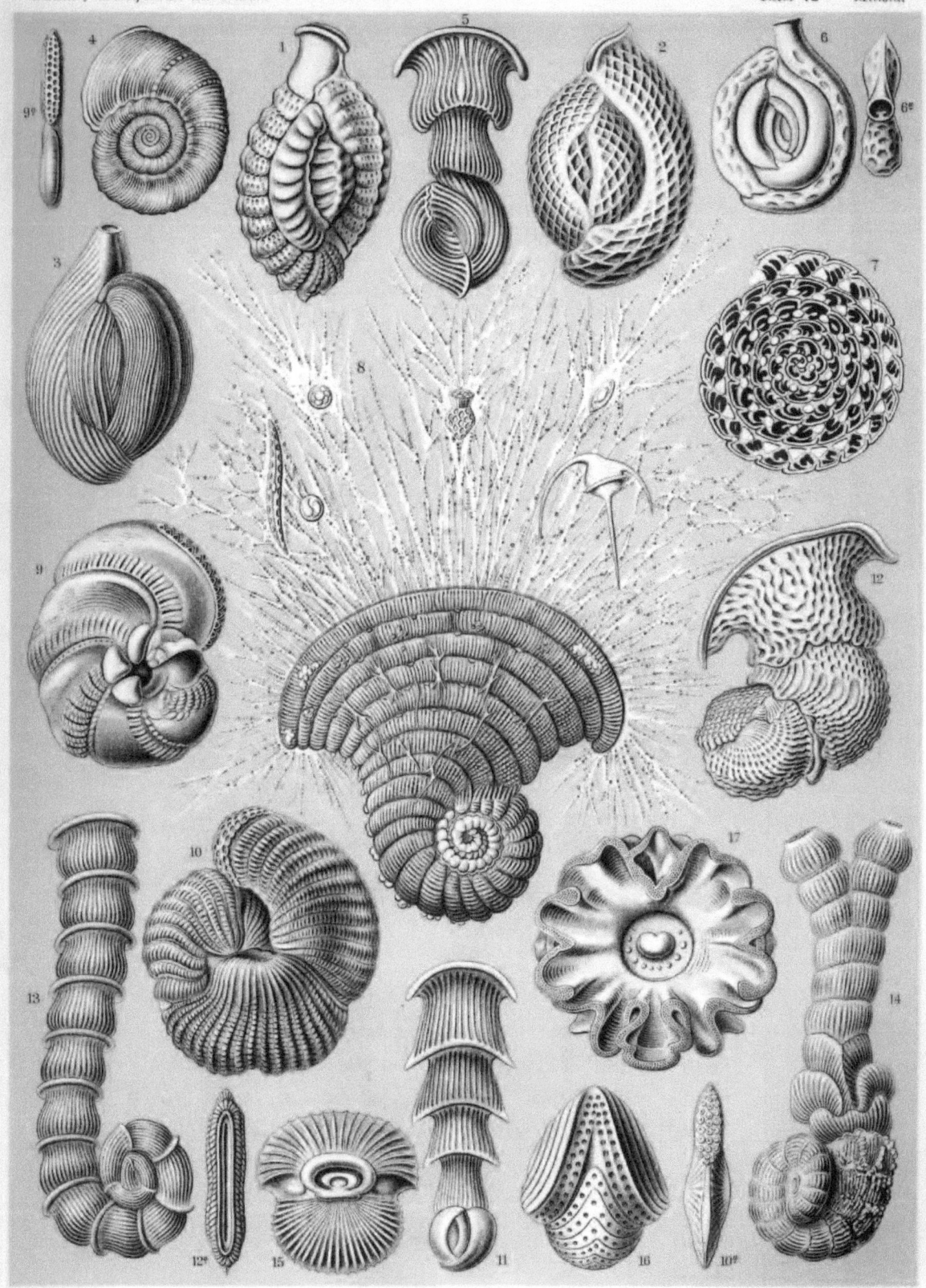

Talamophora. — Kammerlinge.

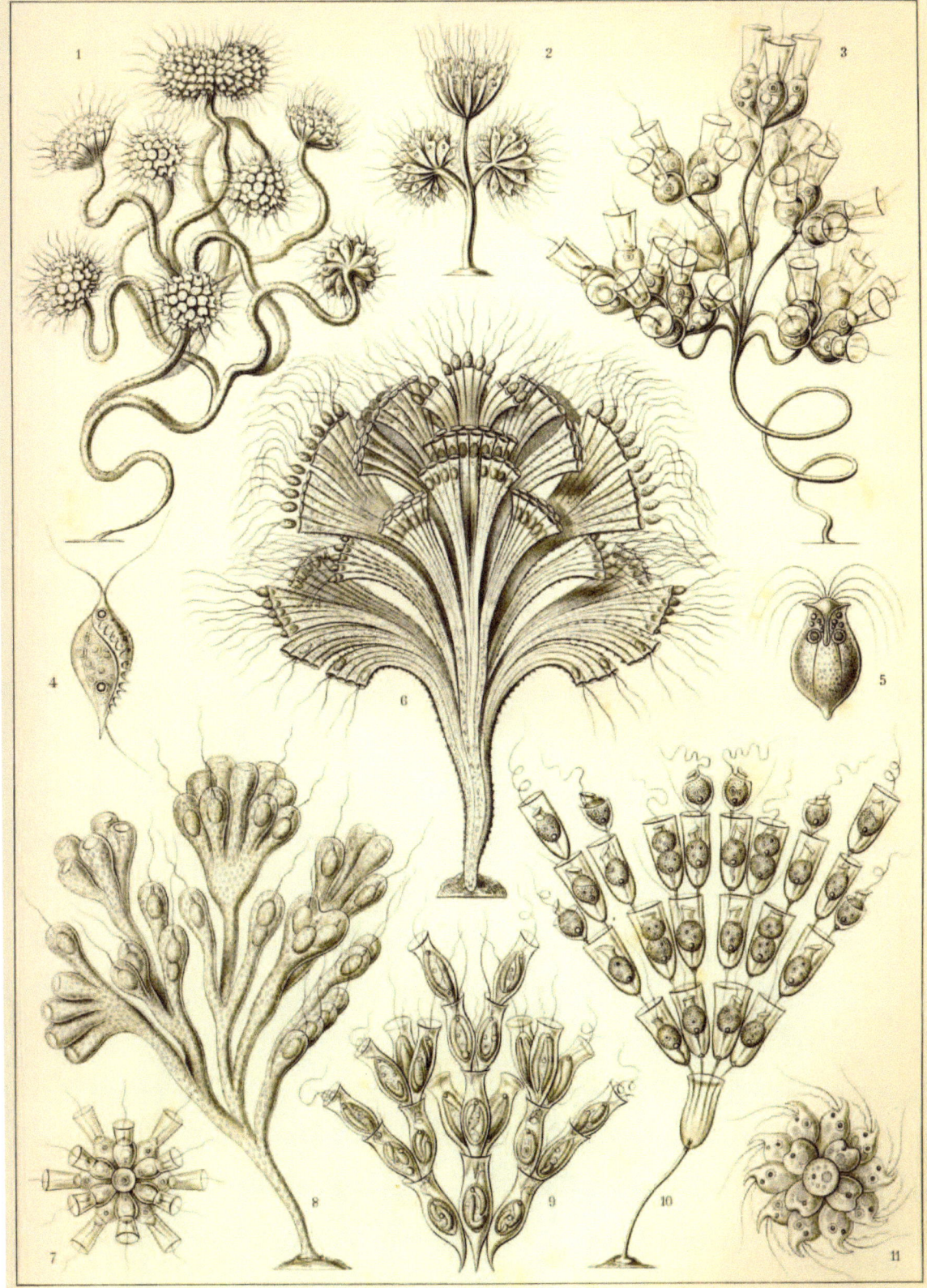

Flagellata. — Geißlinge.

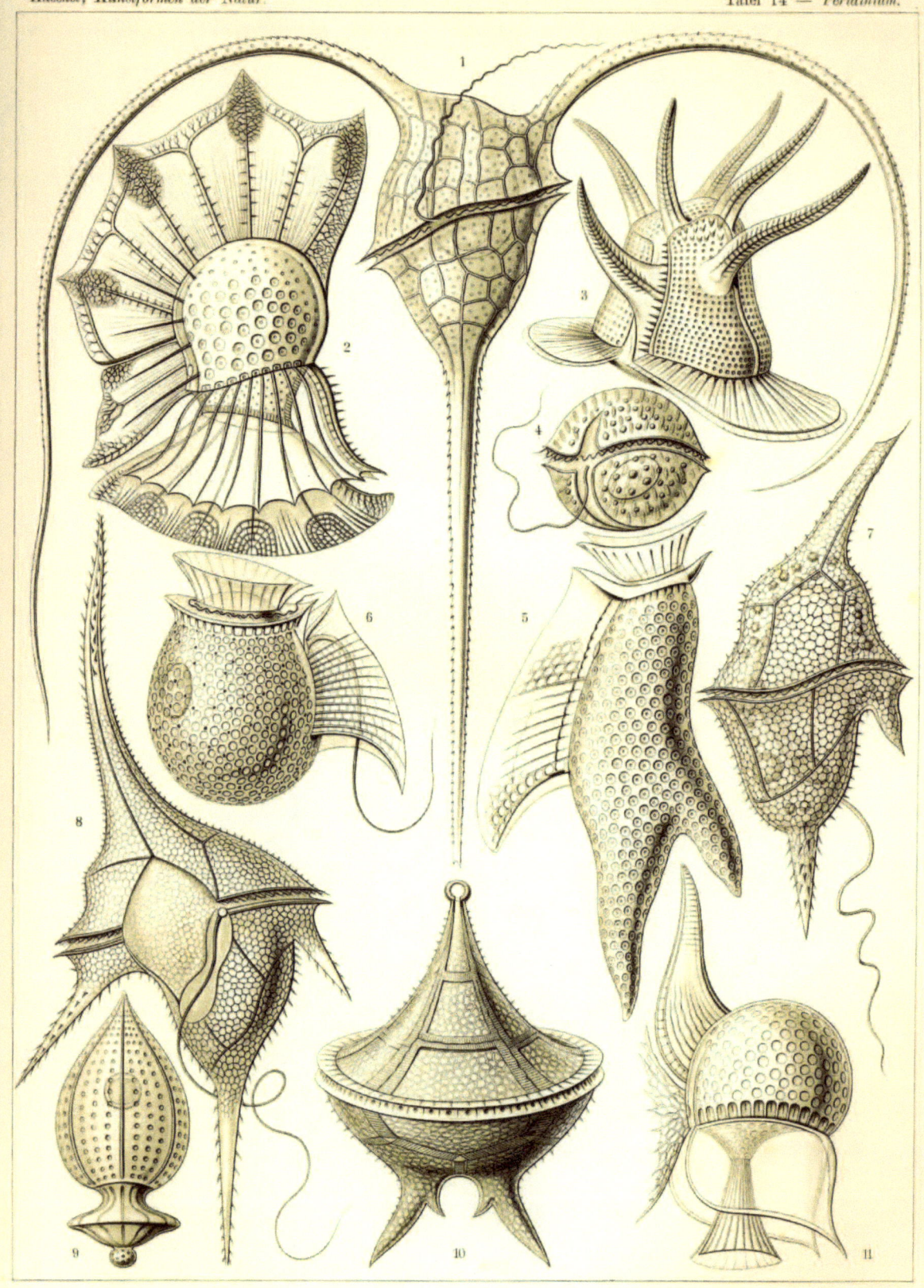

Peridinea. — Geißelhütchen.

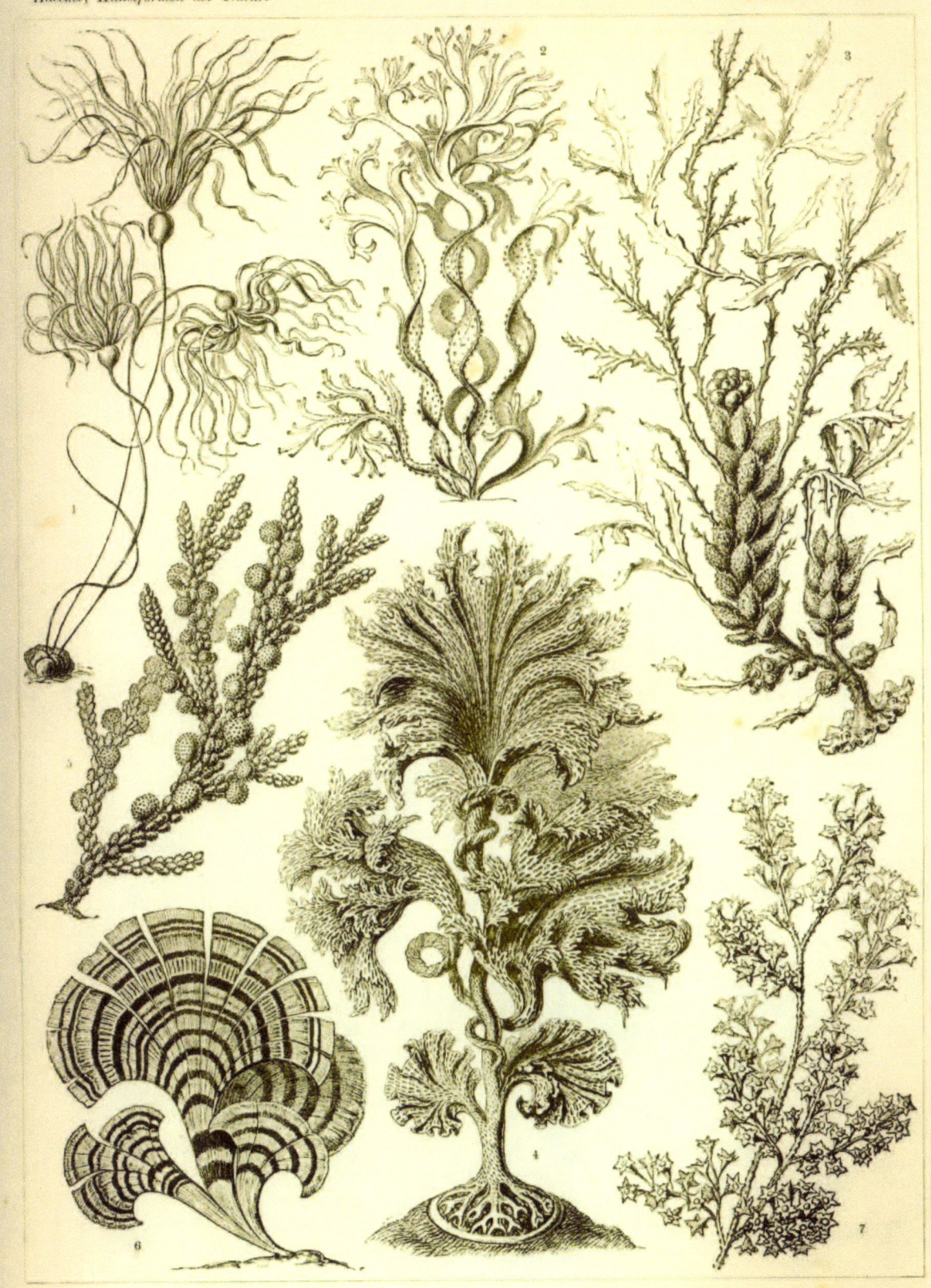

Fucoideae. — Brauntange.

Narcomedusae. — Spangenquallen.

Siphonophorae. — Staatsquallen.

Discomedusae. — Scheibenquallen.

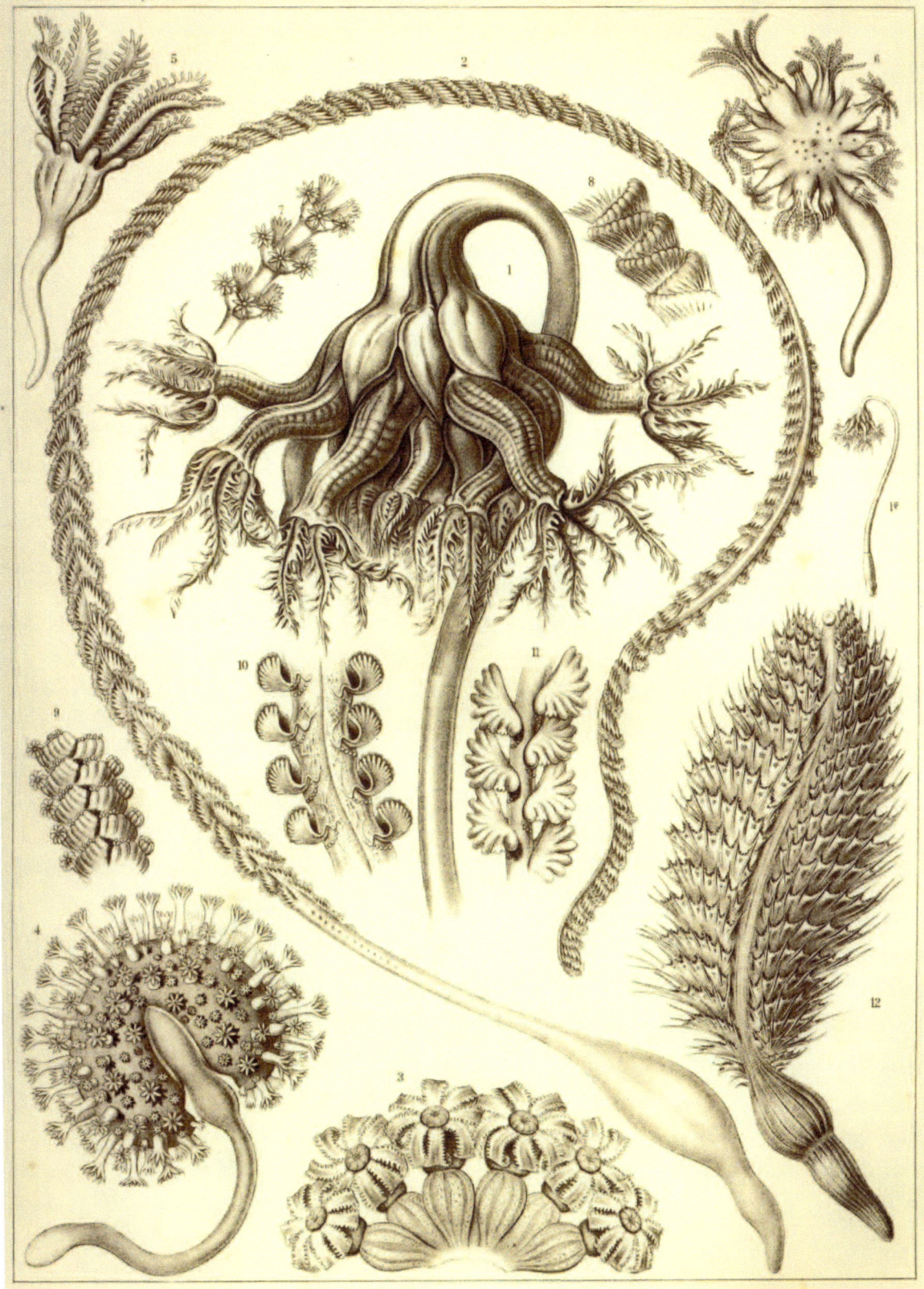

Pennatulida. — Federkorallen.

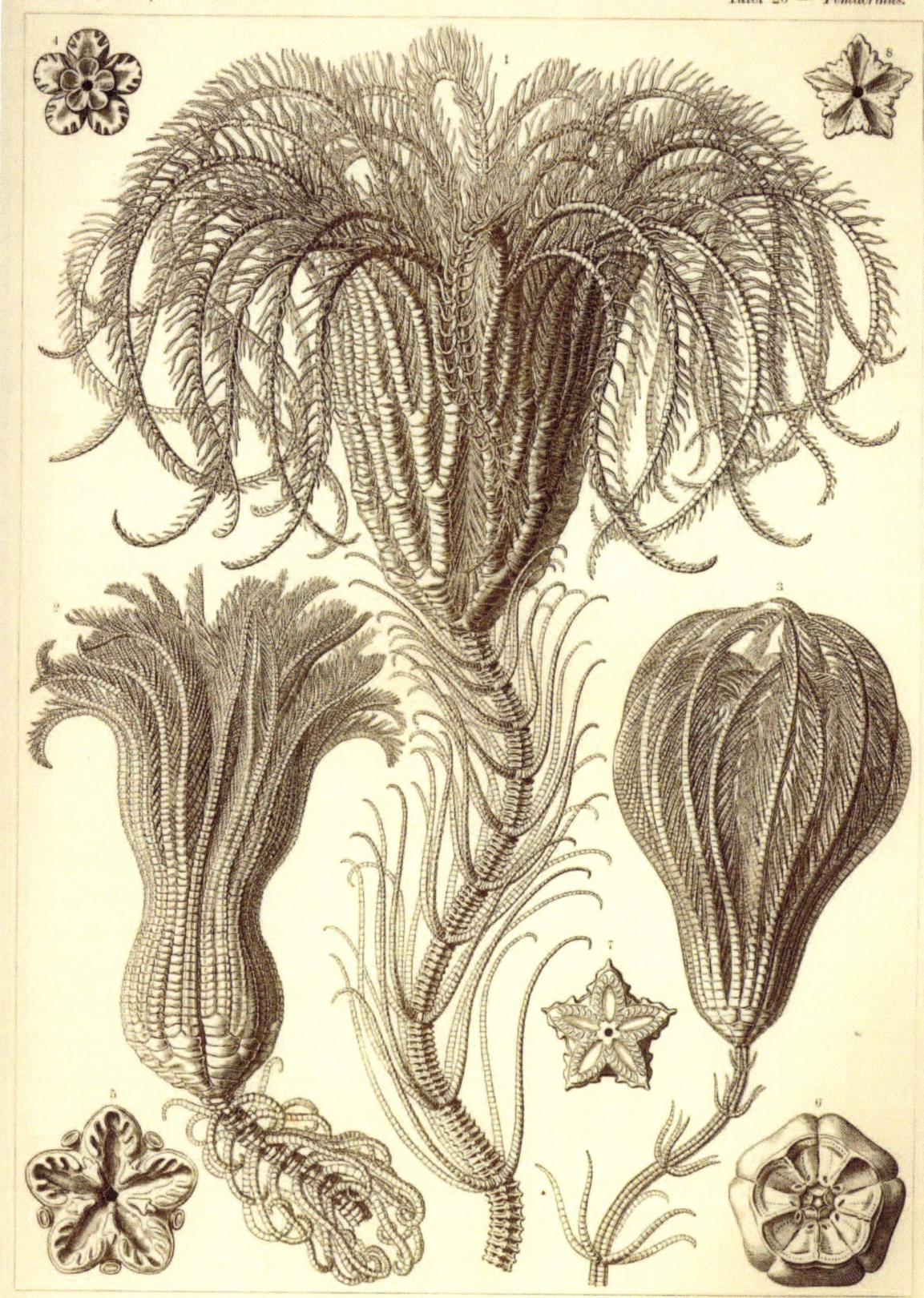

Crinoidea. — Palmensterne.

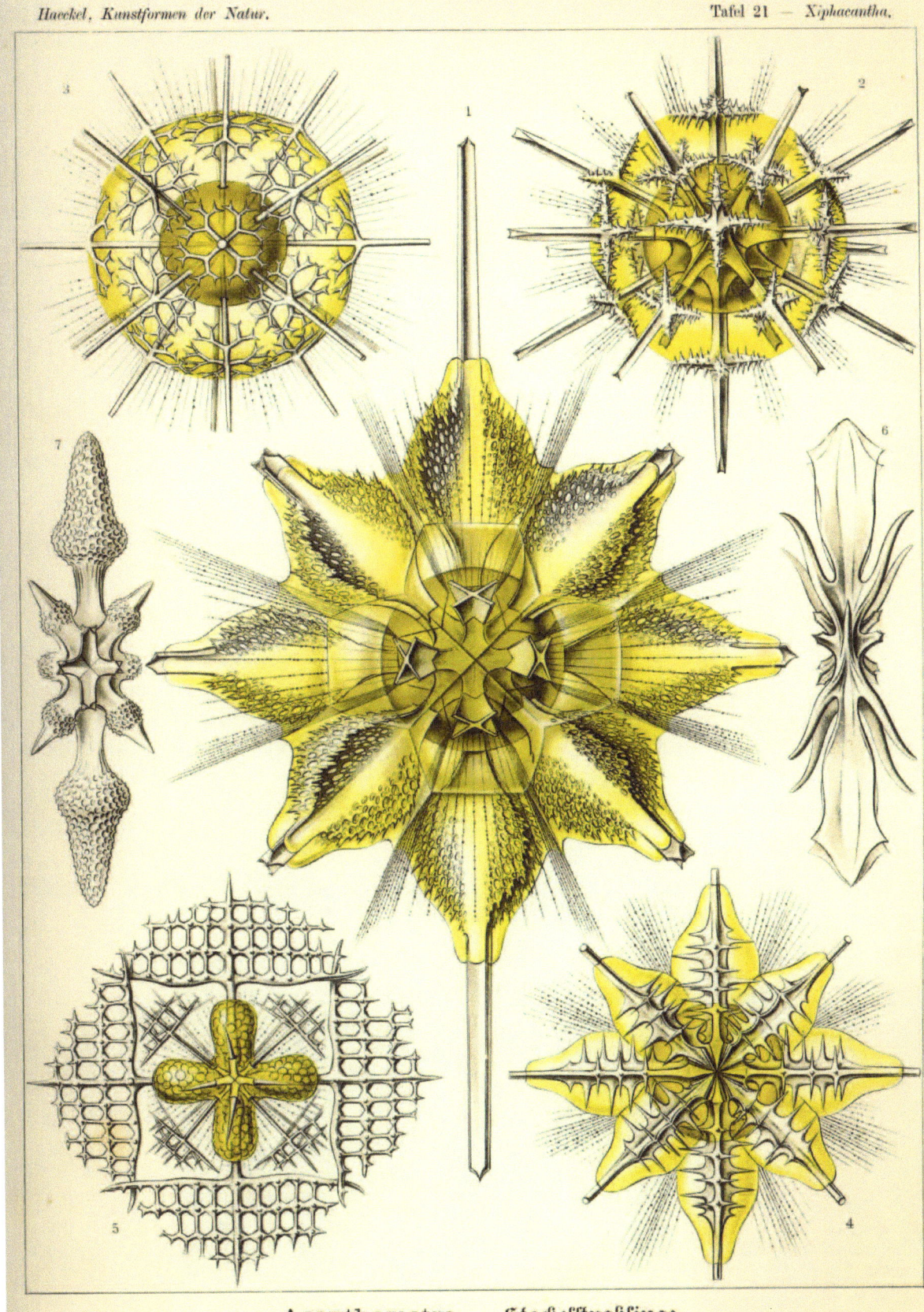

Acanthometra. — Stachelstrahlinge.

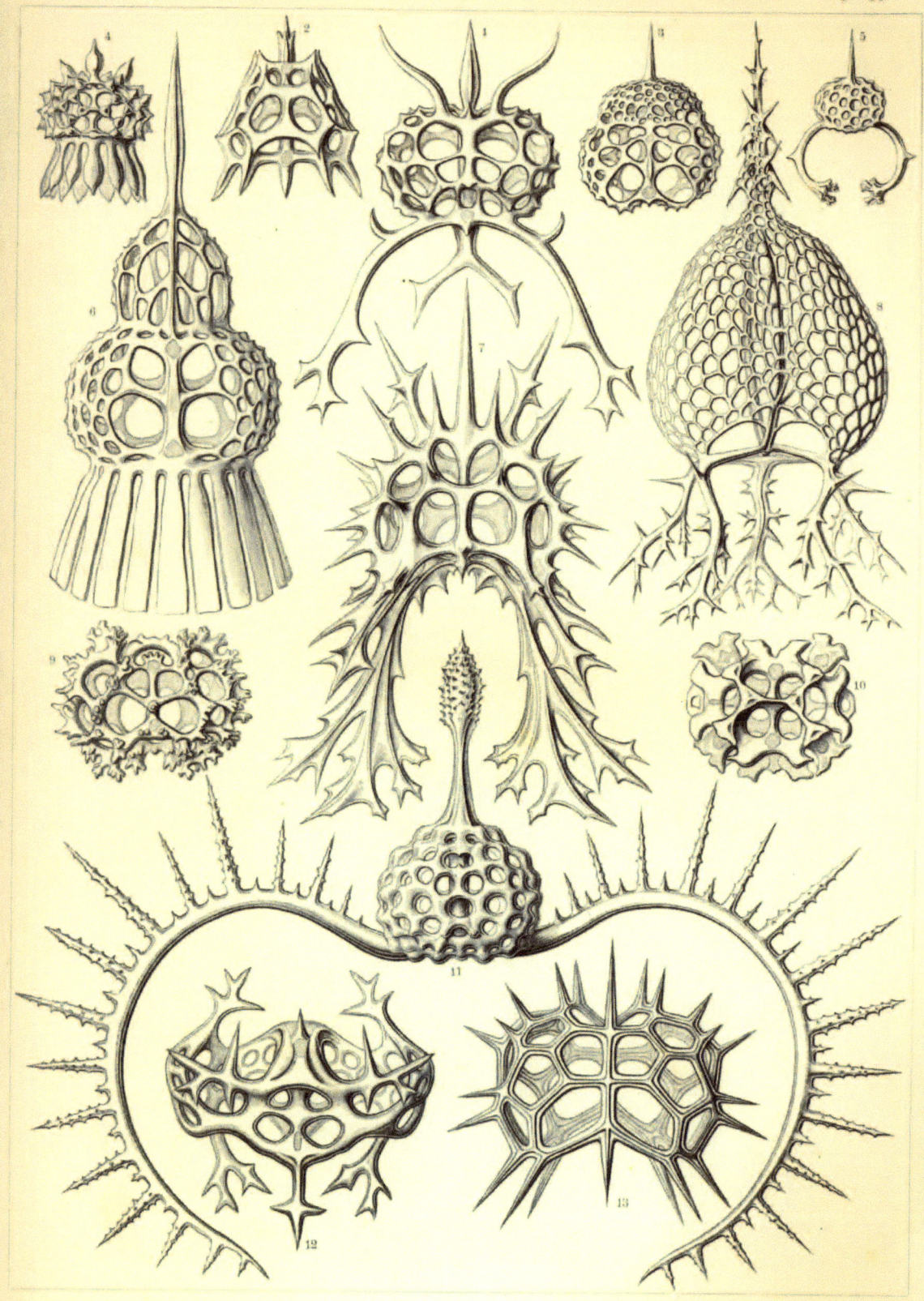

Spyroidea. — Rüsschenstrahlinge.

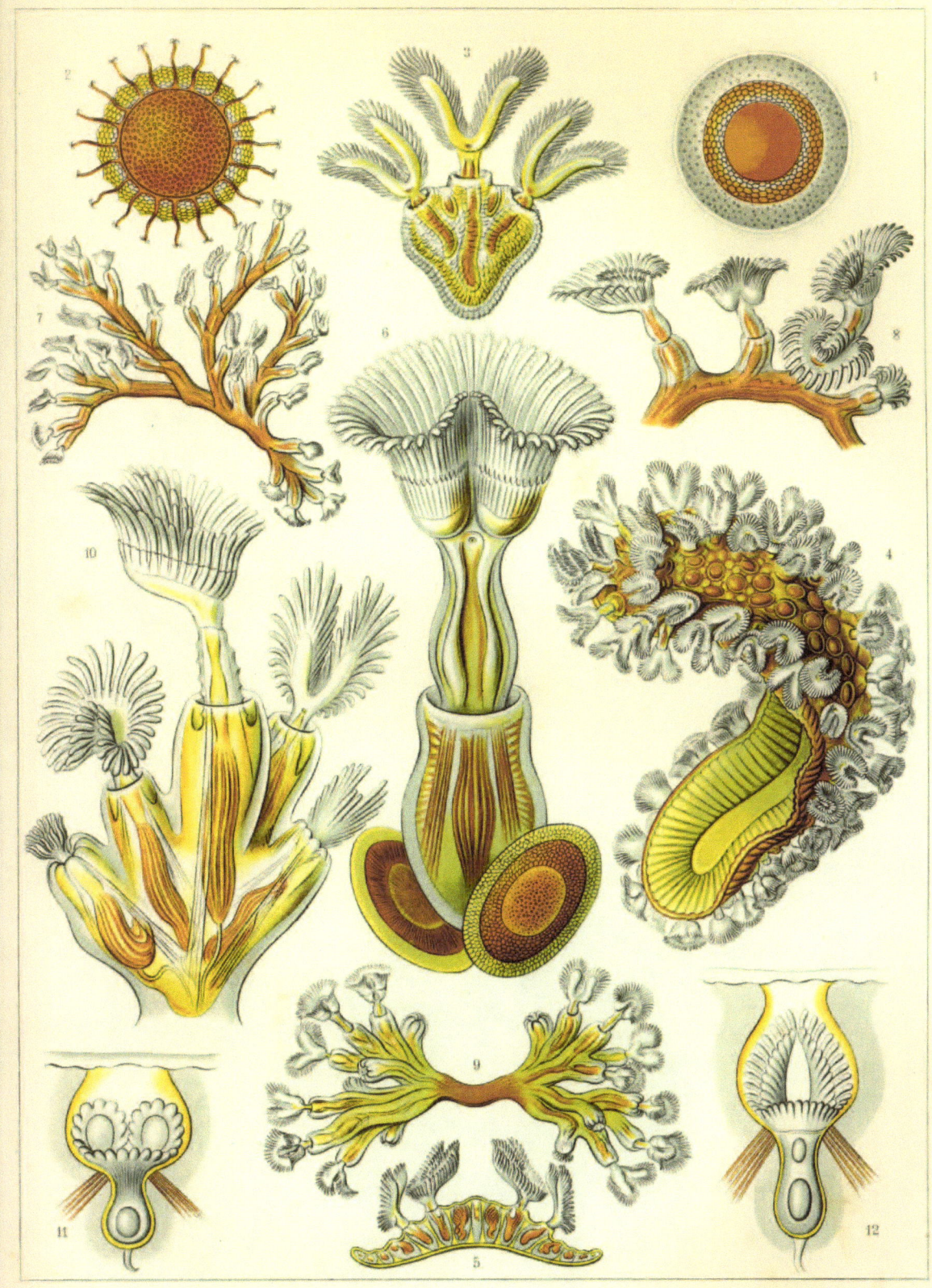

Bryozoa. — Moostiere.

Desmidiea. — Zierdinge.

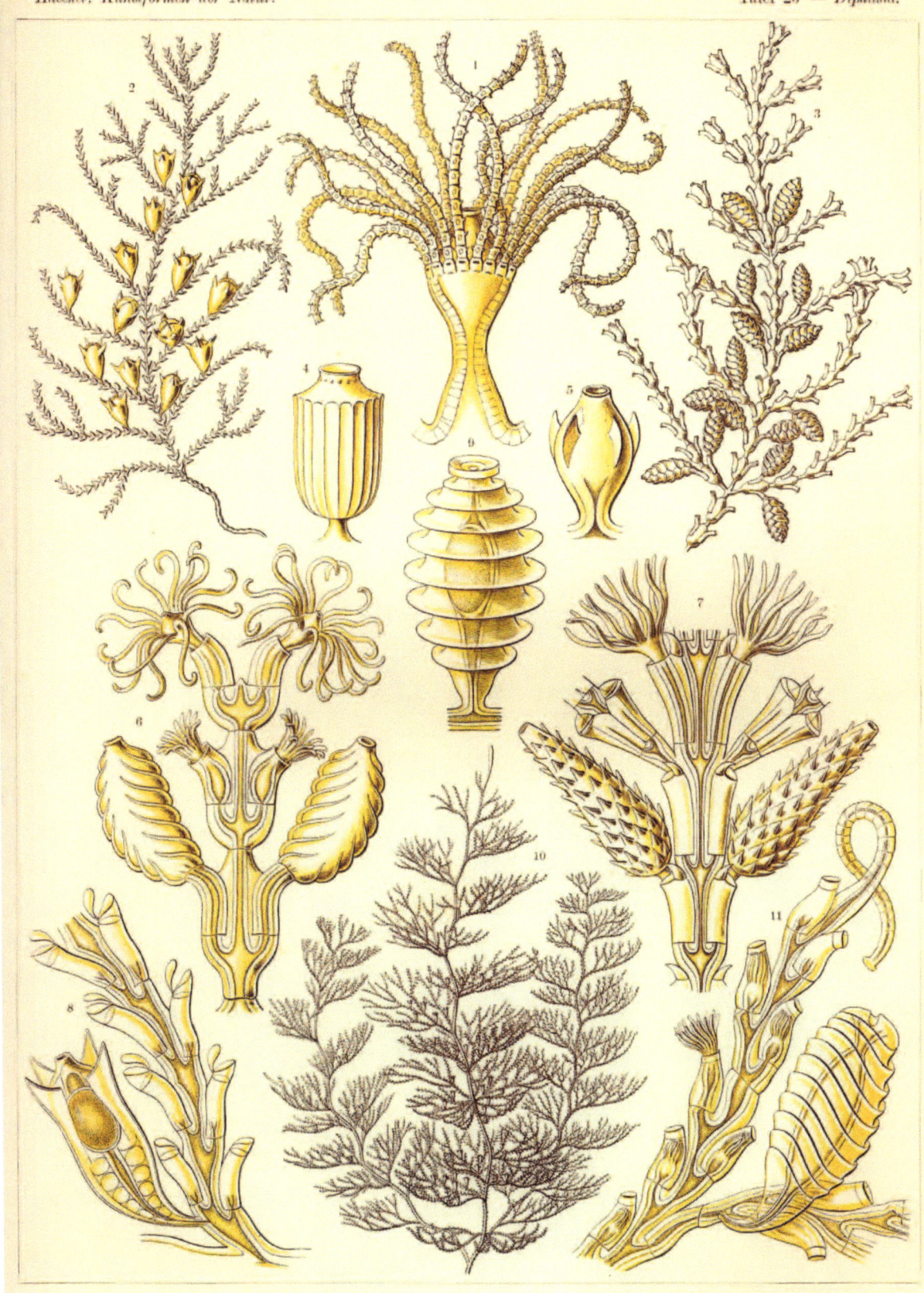

Sertulariae. — Reihenpolypen.

Trachomedusae. — Kolbenquallen.

Ctenophorae. — Kammquallen.

Discomedusae. — Scheibenquallen.

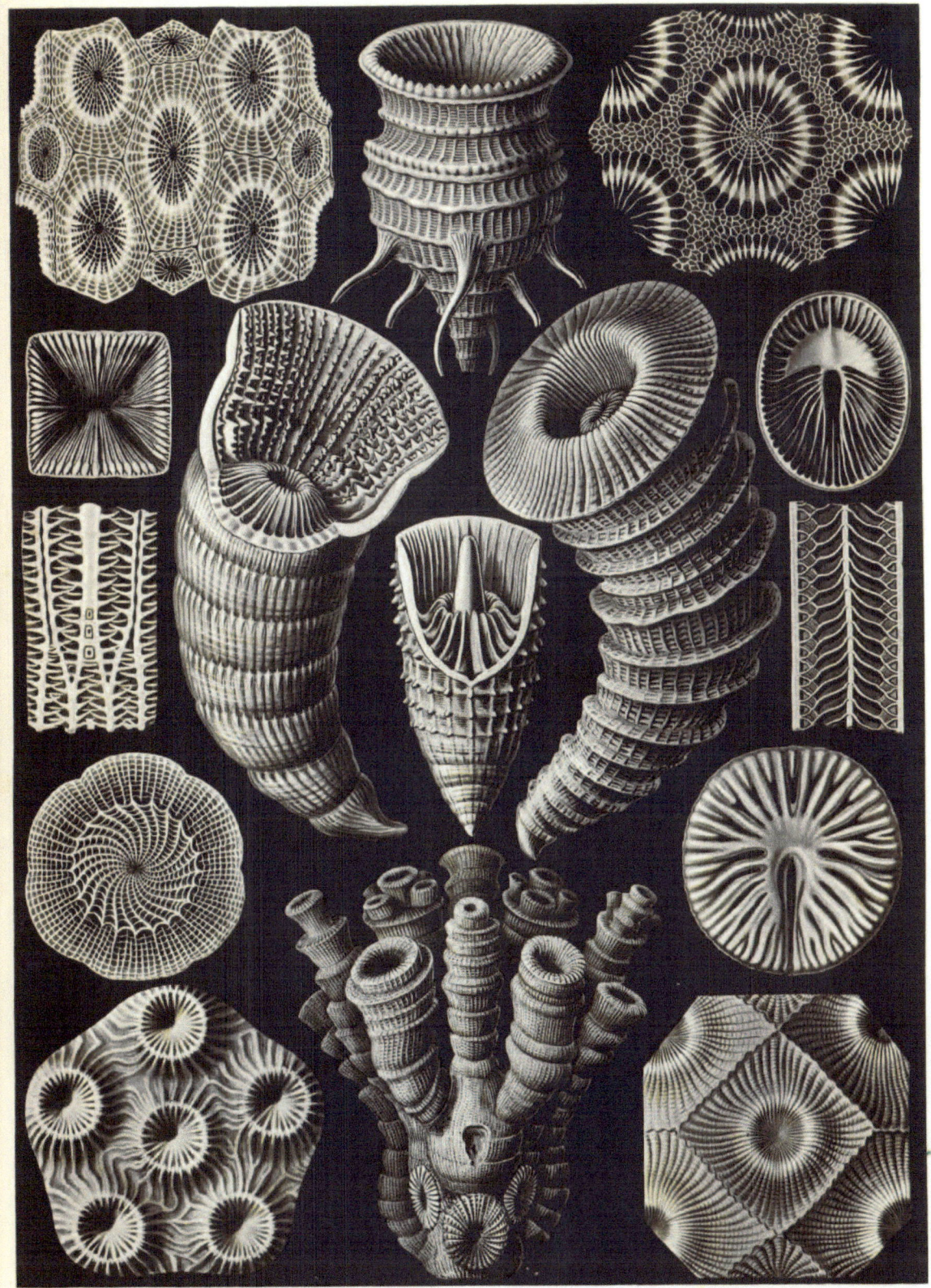

Tetracoralla. — Vierstrahlige Sternkorallen.

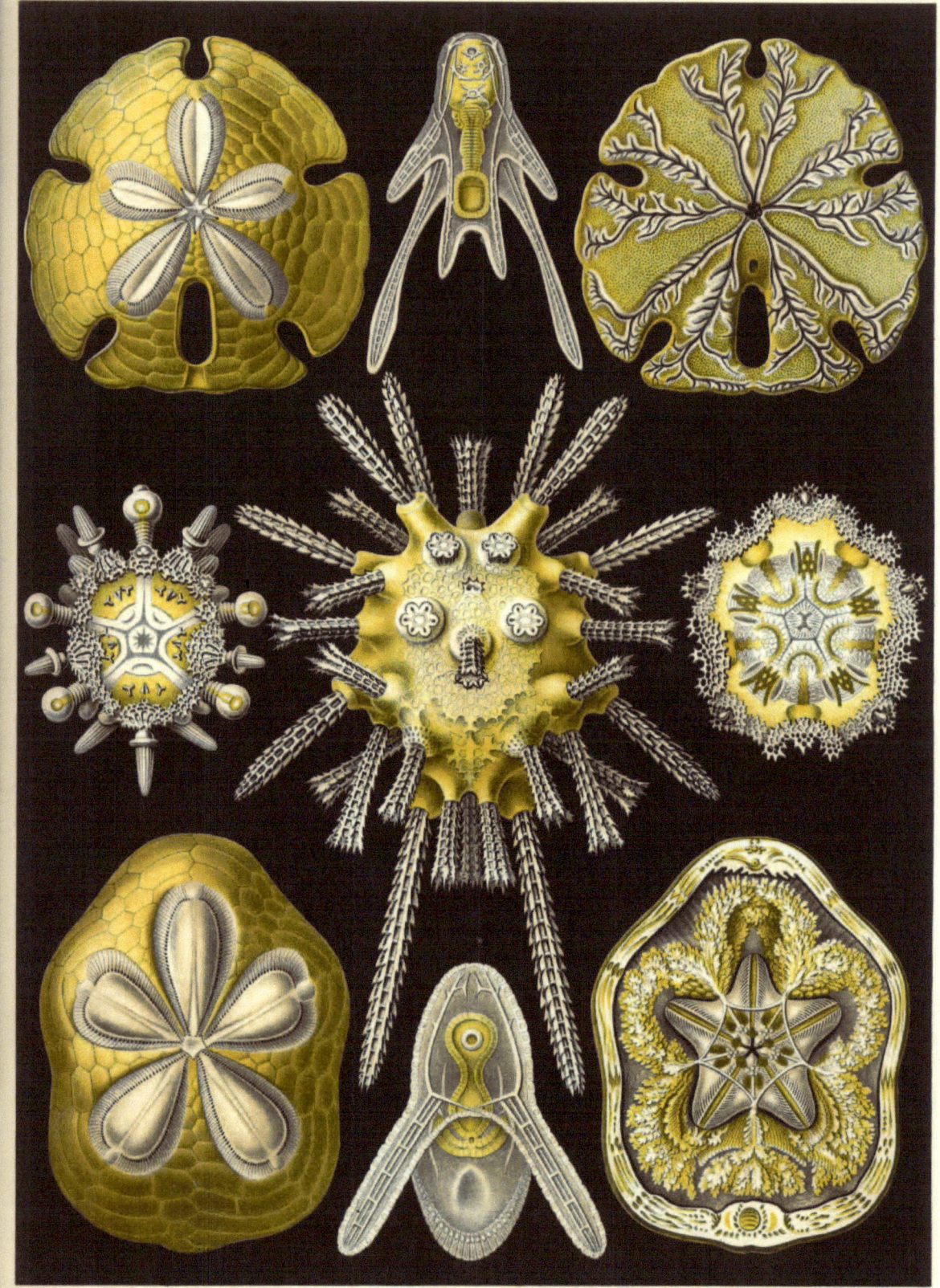

Echinidea. — Igelsterne.

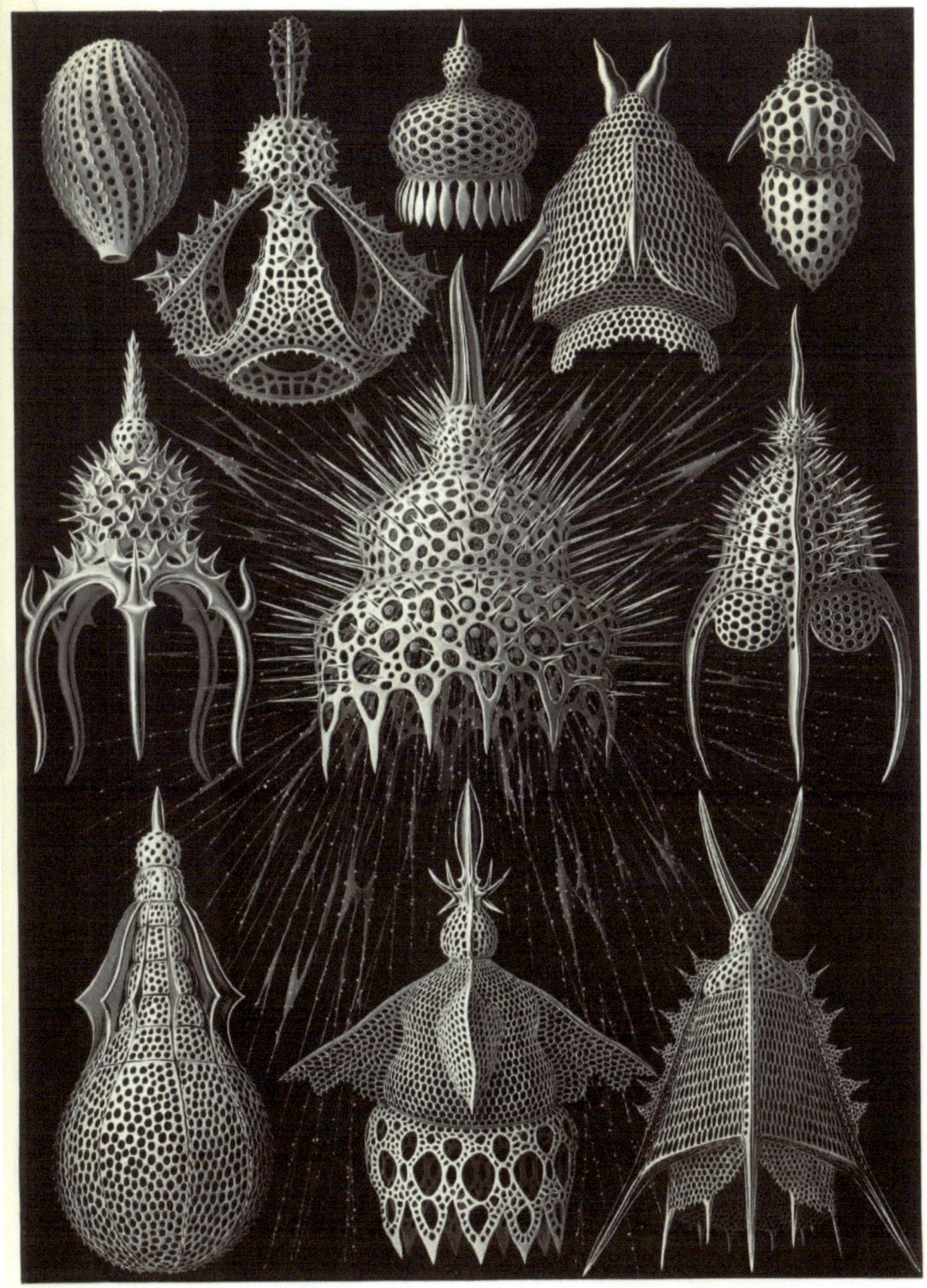

Cyrtoidea. — Flaschenstrahlinge.

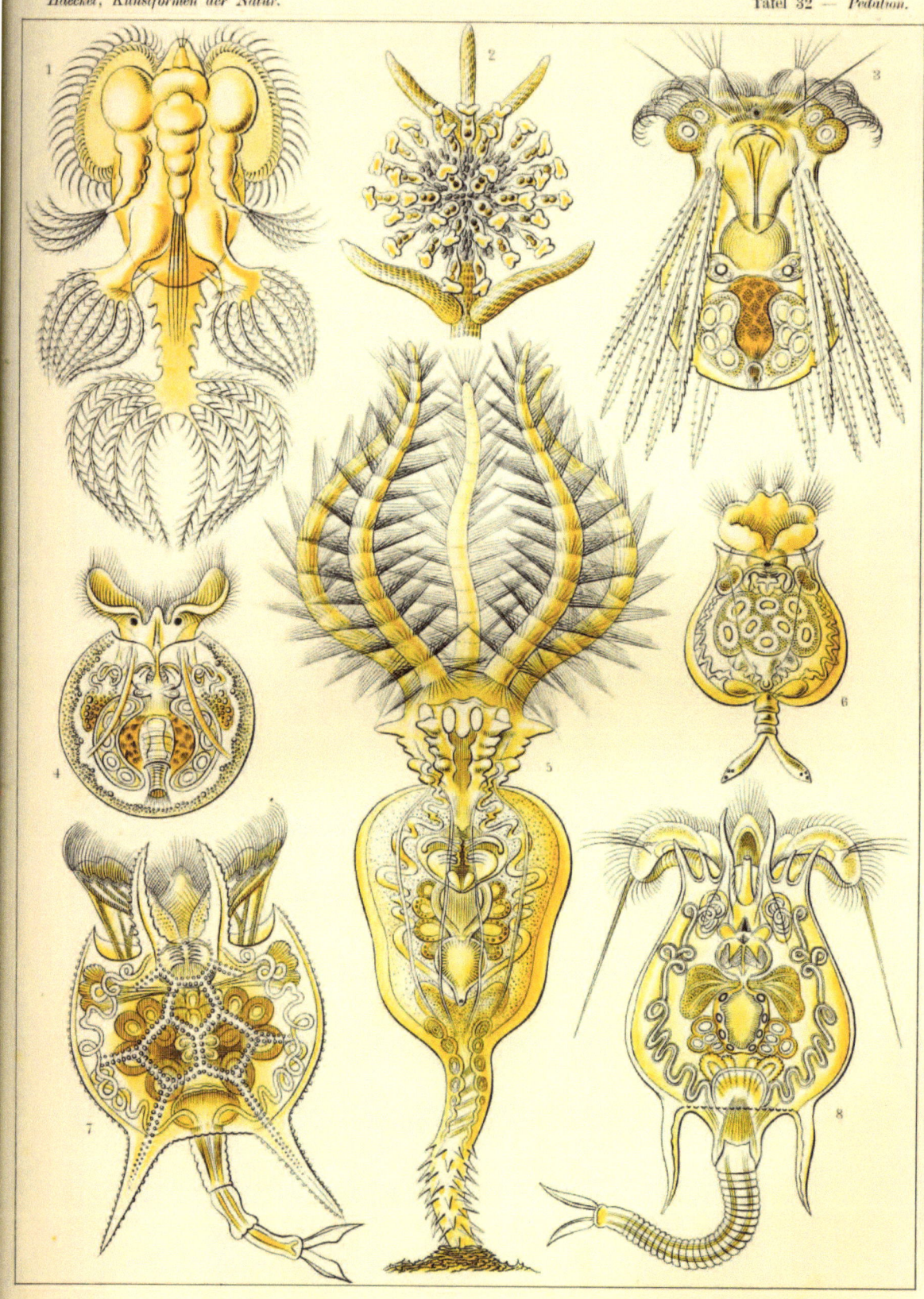

Rotatoria. — Rädertiere.

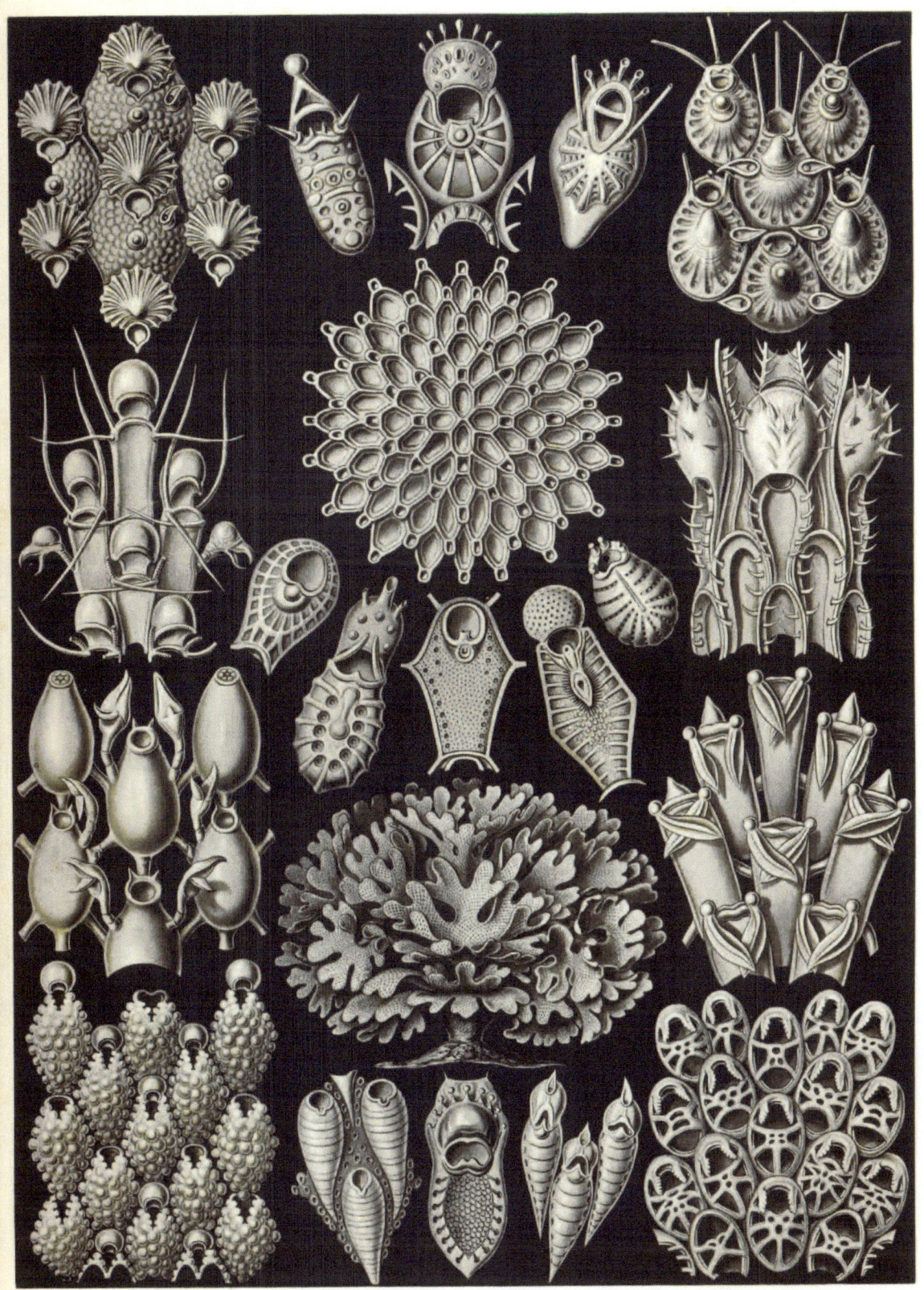

Bryozoa. — Moostiere.

Melethallia. — Gesellige Algetten.

Hexactinellae. — Glasschwämme.

Leptomedusae. — Faltenquallen.

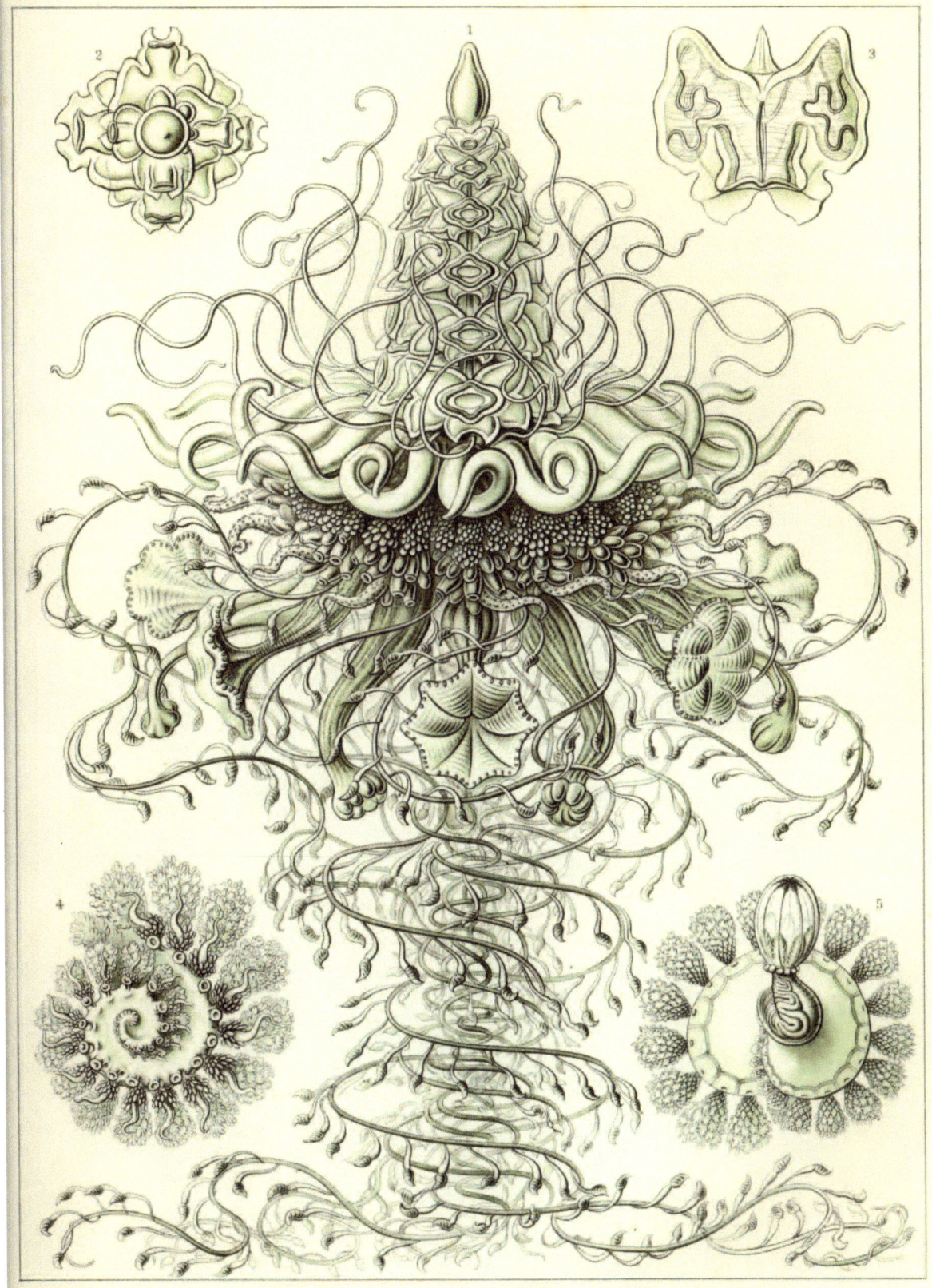

Siphonophorae. — Staatsquallen.

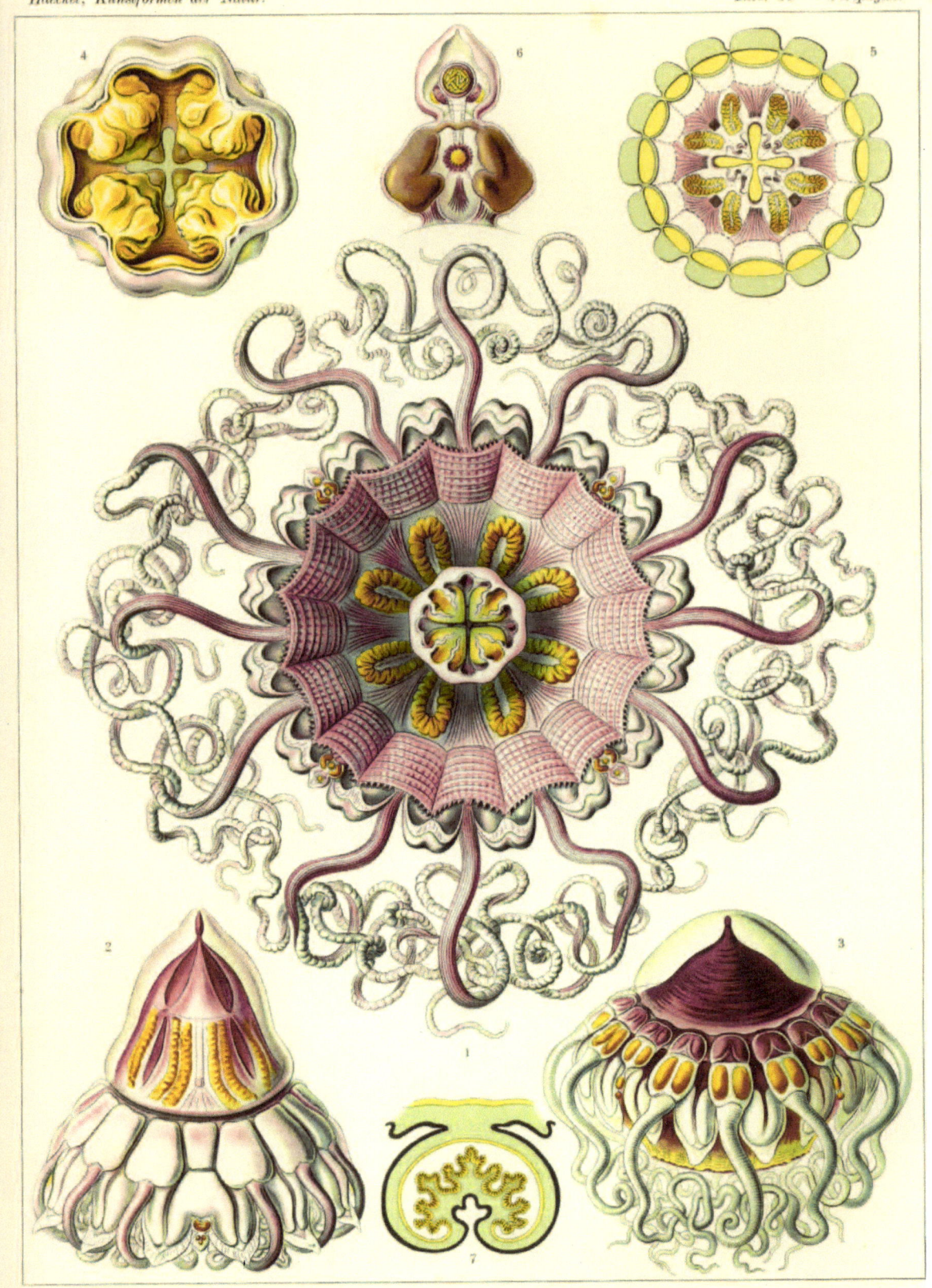

Peromedusae. — Taſchenquallen.

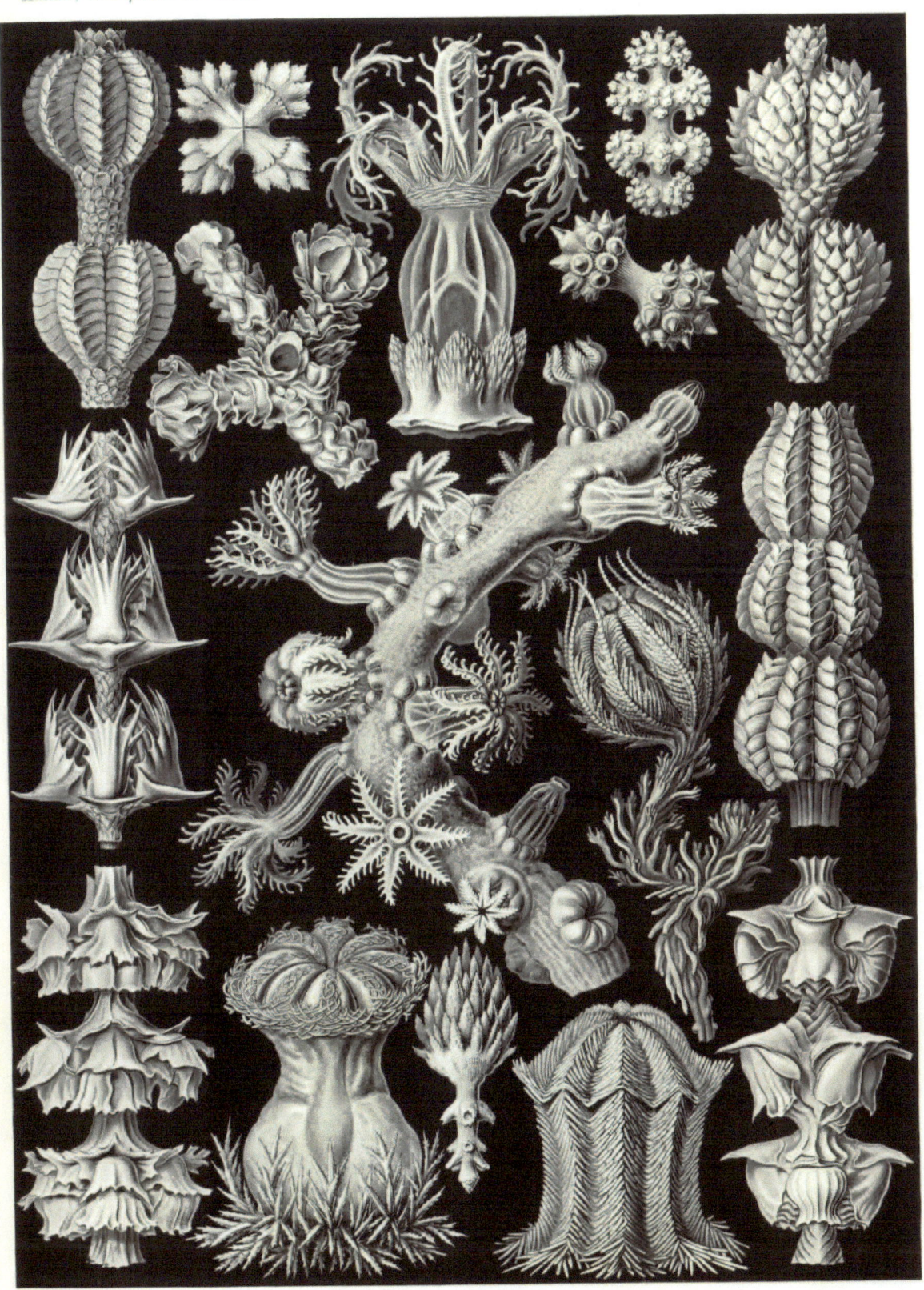

Gorgonida. — Rindenkorallen.

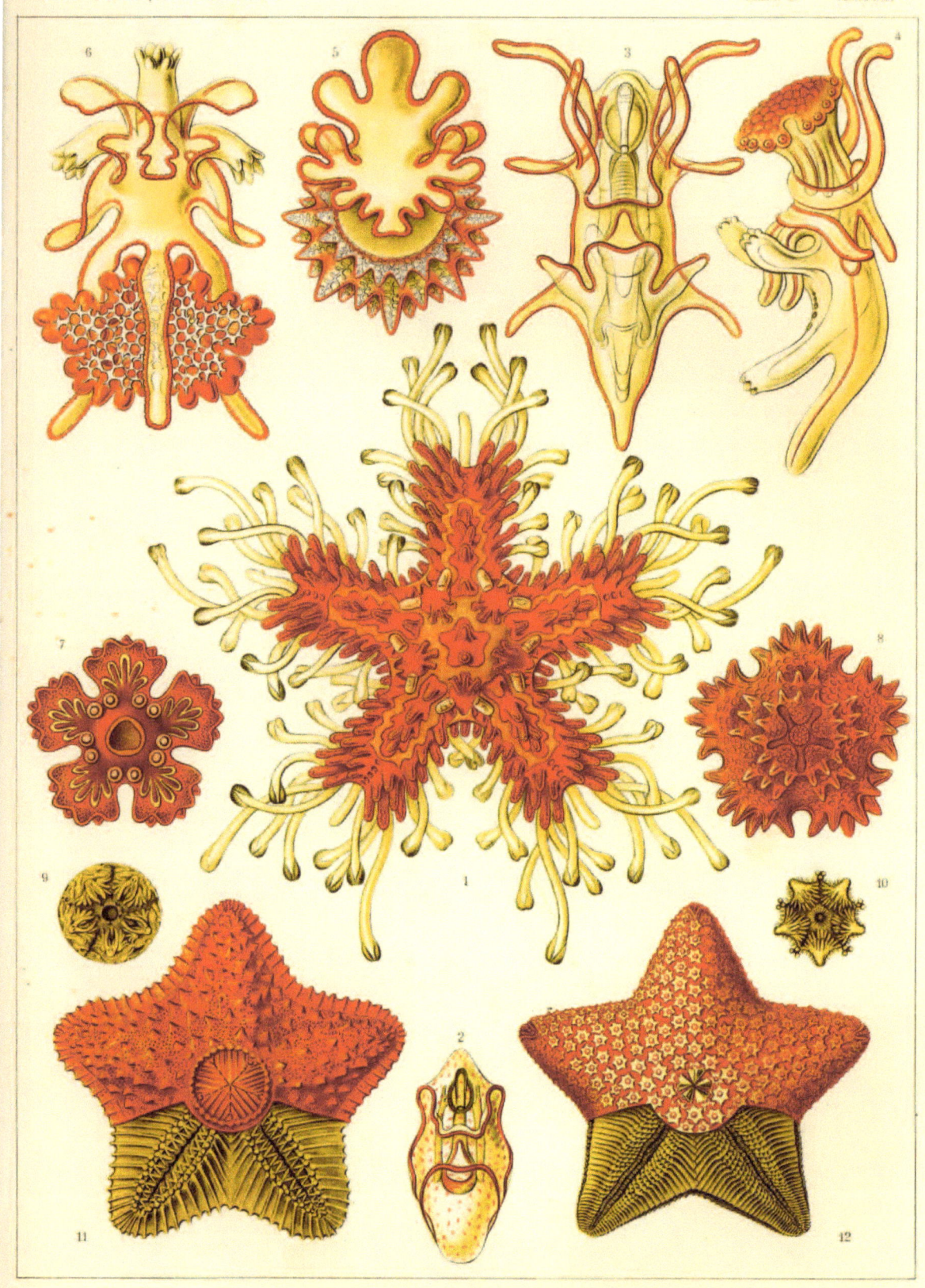

Asteridea. — Seesterne.

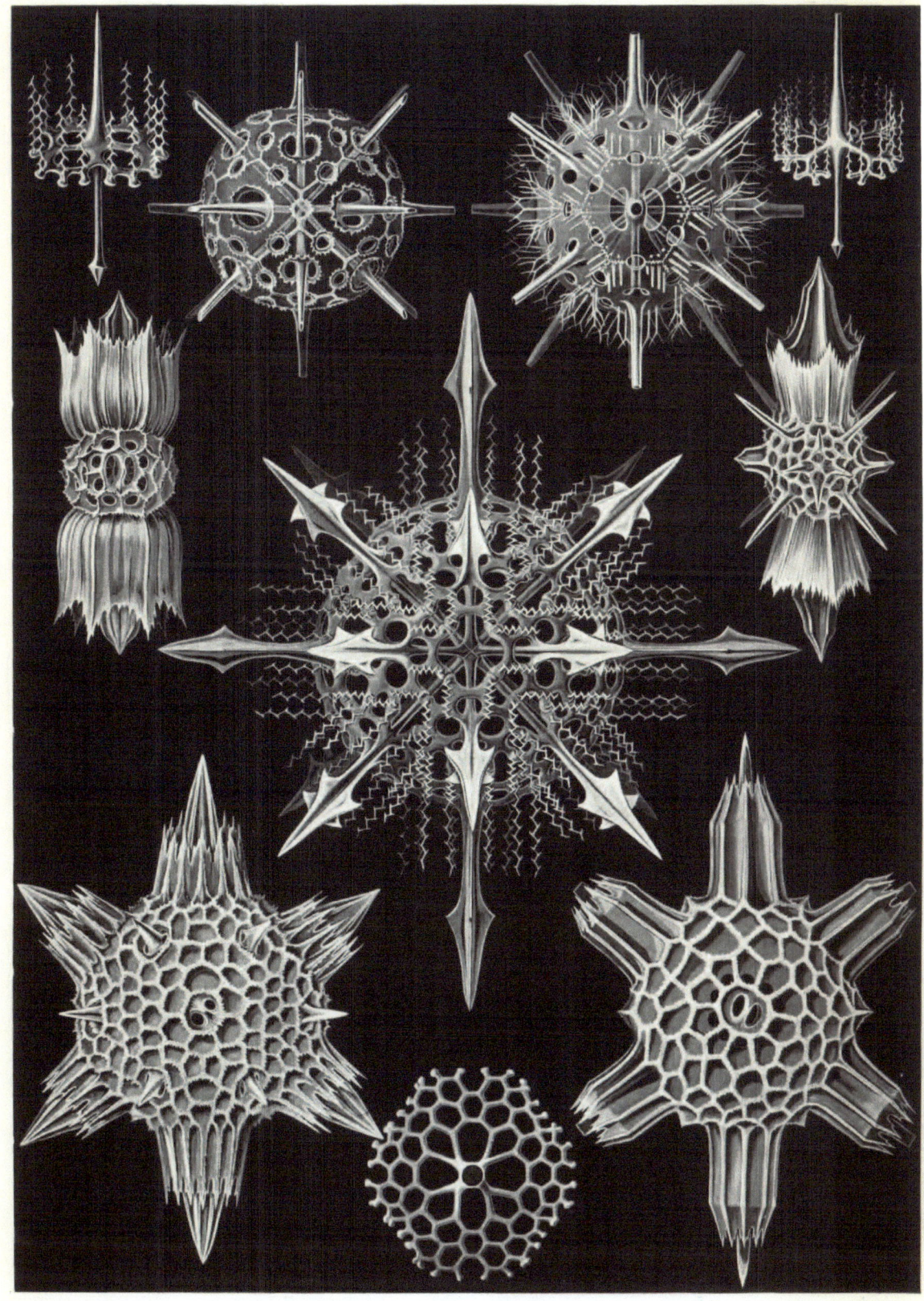

Acanthophracta. — Wunderstrahlinge.

Ostraciontes. — Kofferfische.

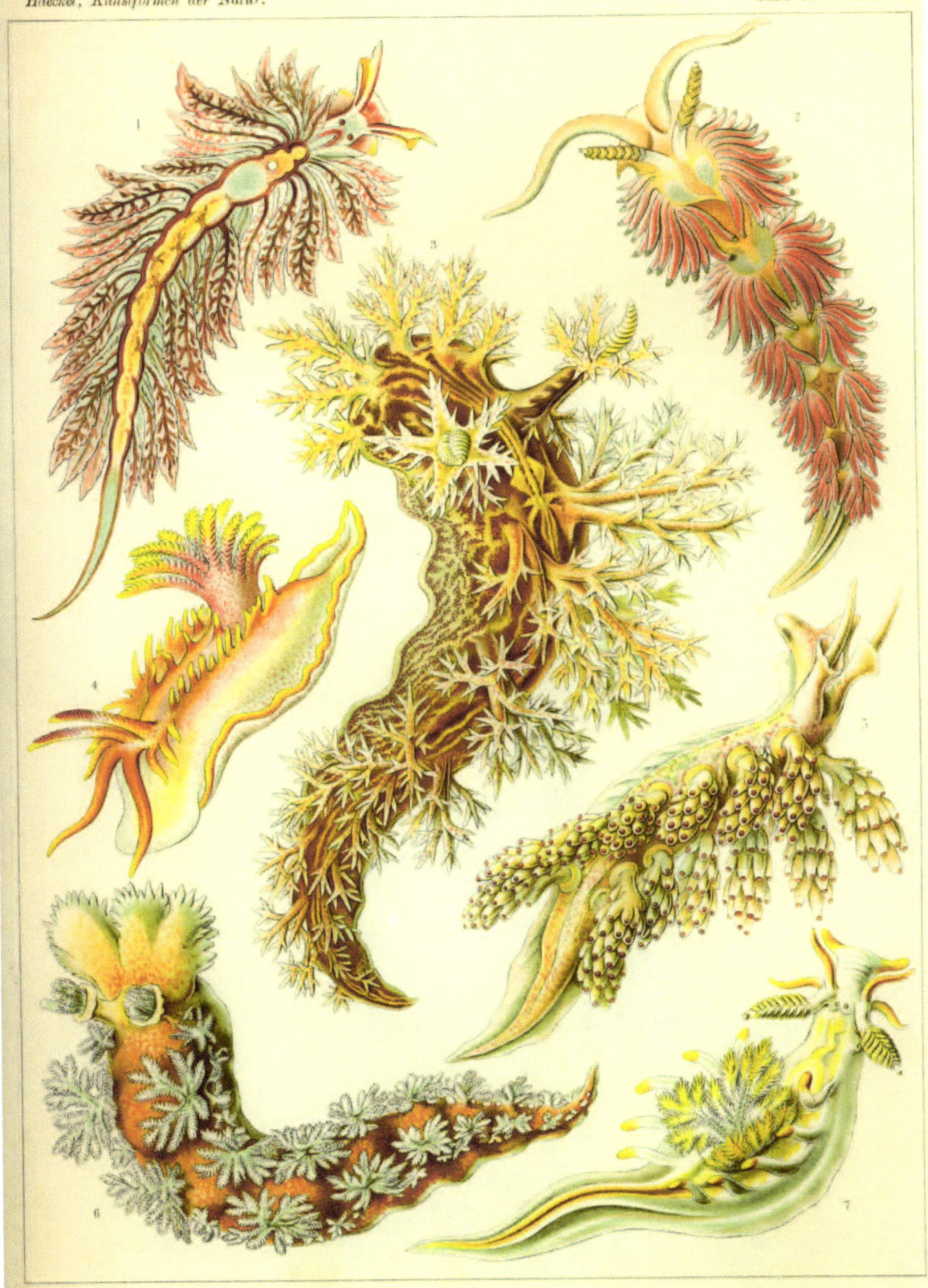

Nudibranchia. — Nacktkiemen-Schnecken.

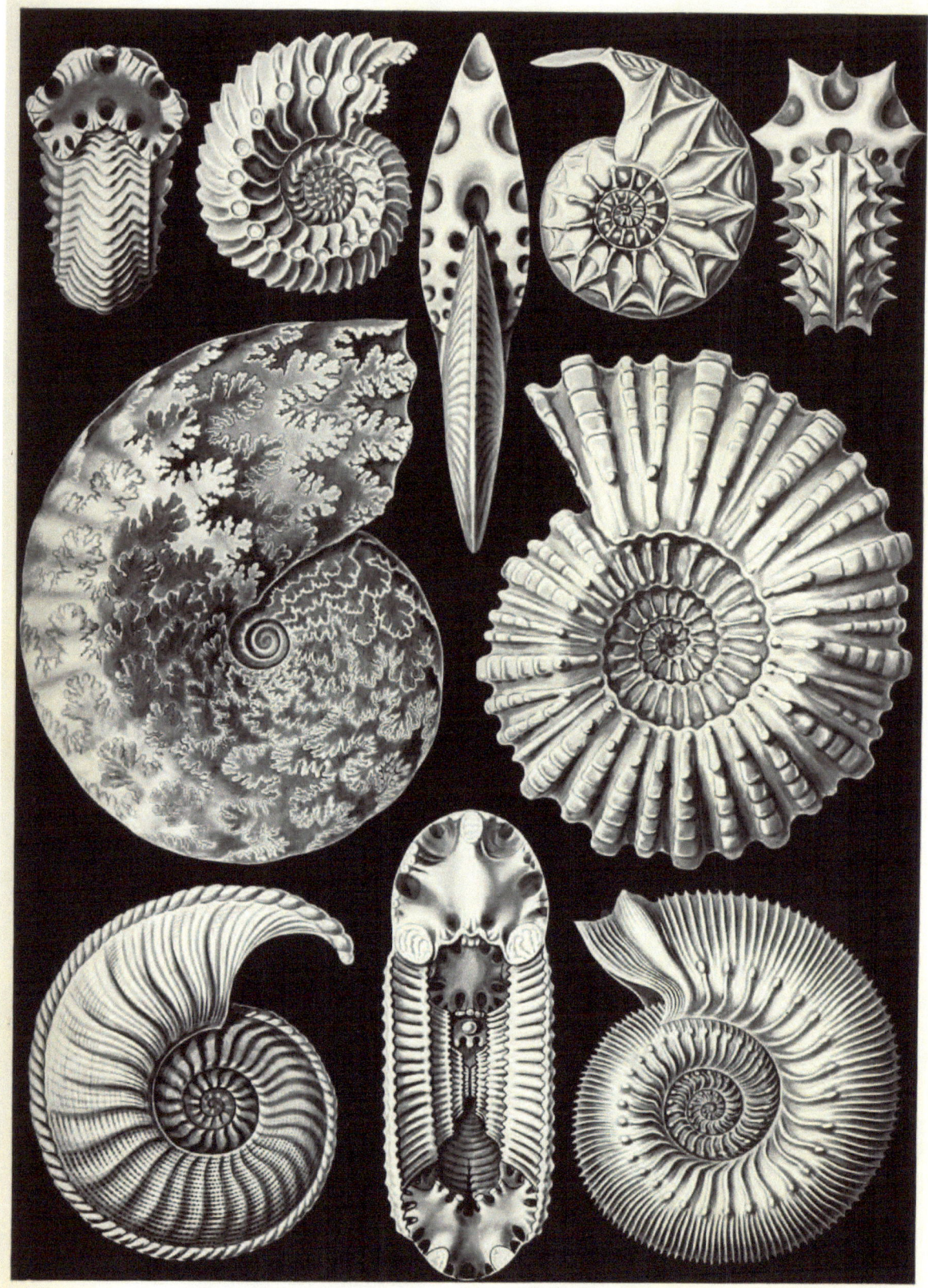

Ammonitida. — Ammonshörner.

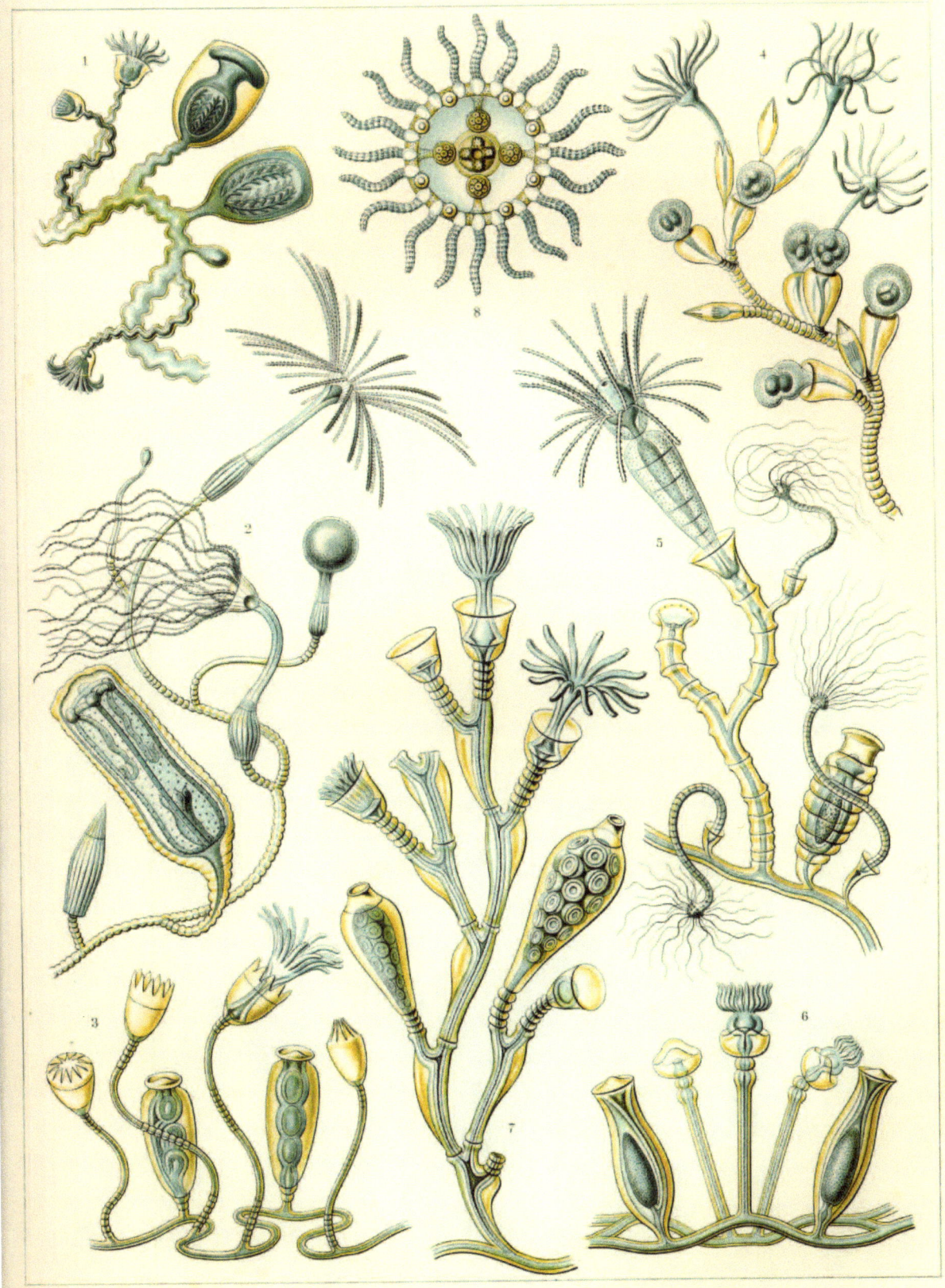

Campanariae. — Glockenpolypen.

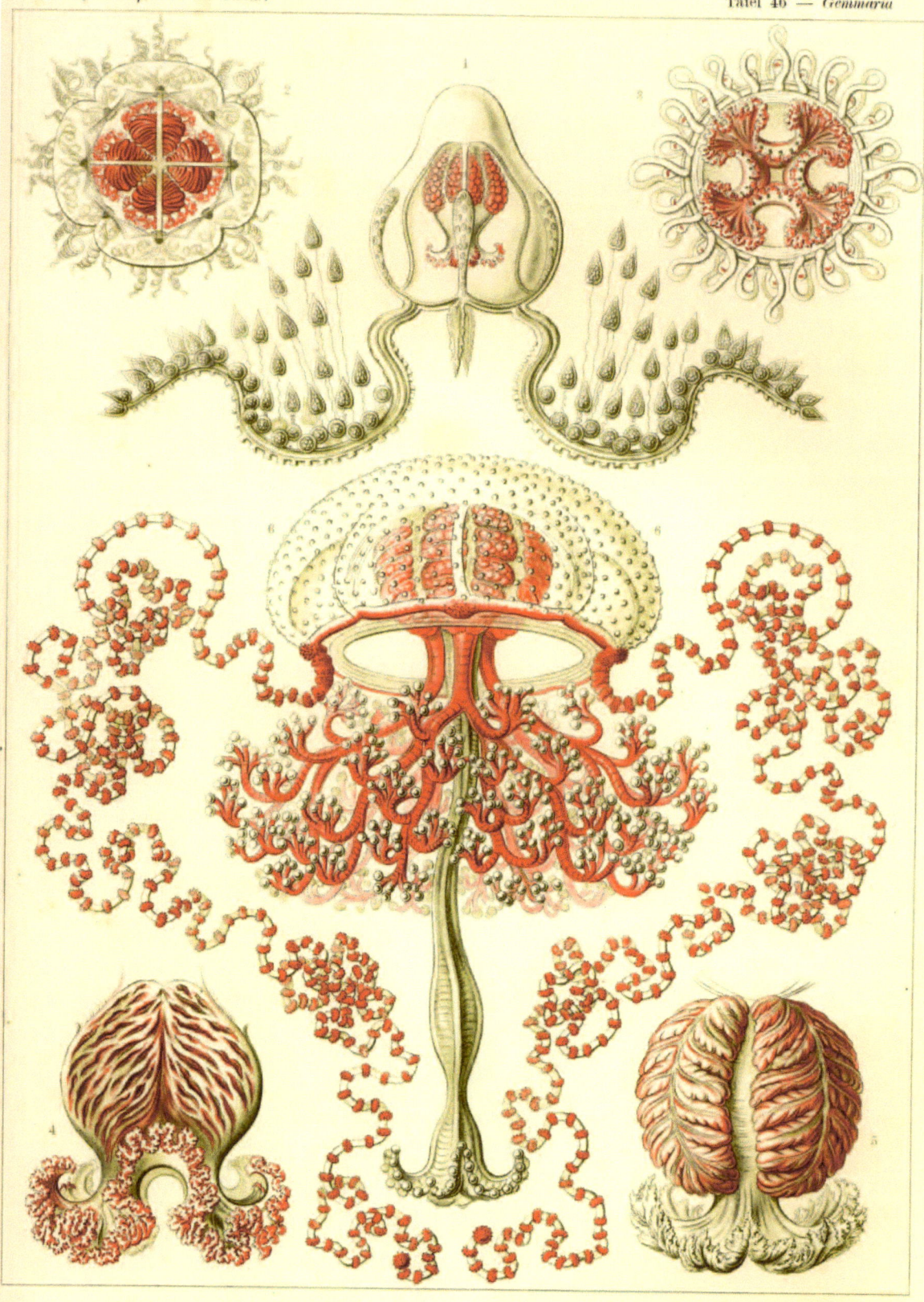

Anthomedusae. — Blumenquallen.

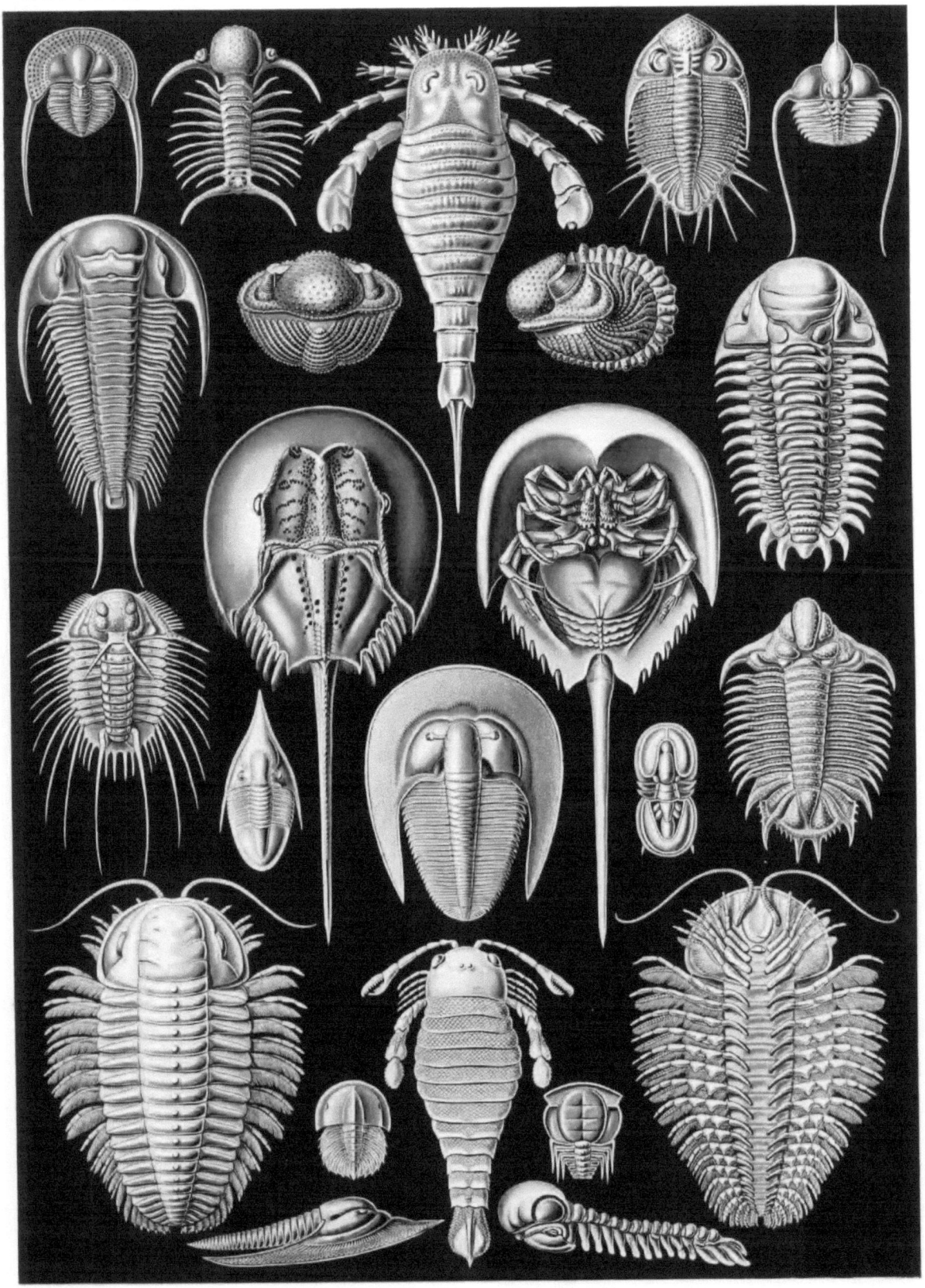

Aspidonia. — Schildtiere.

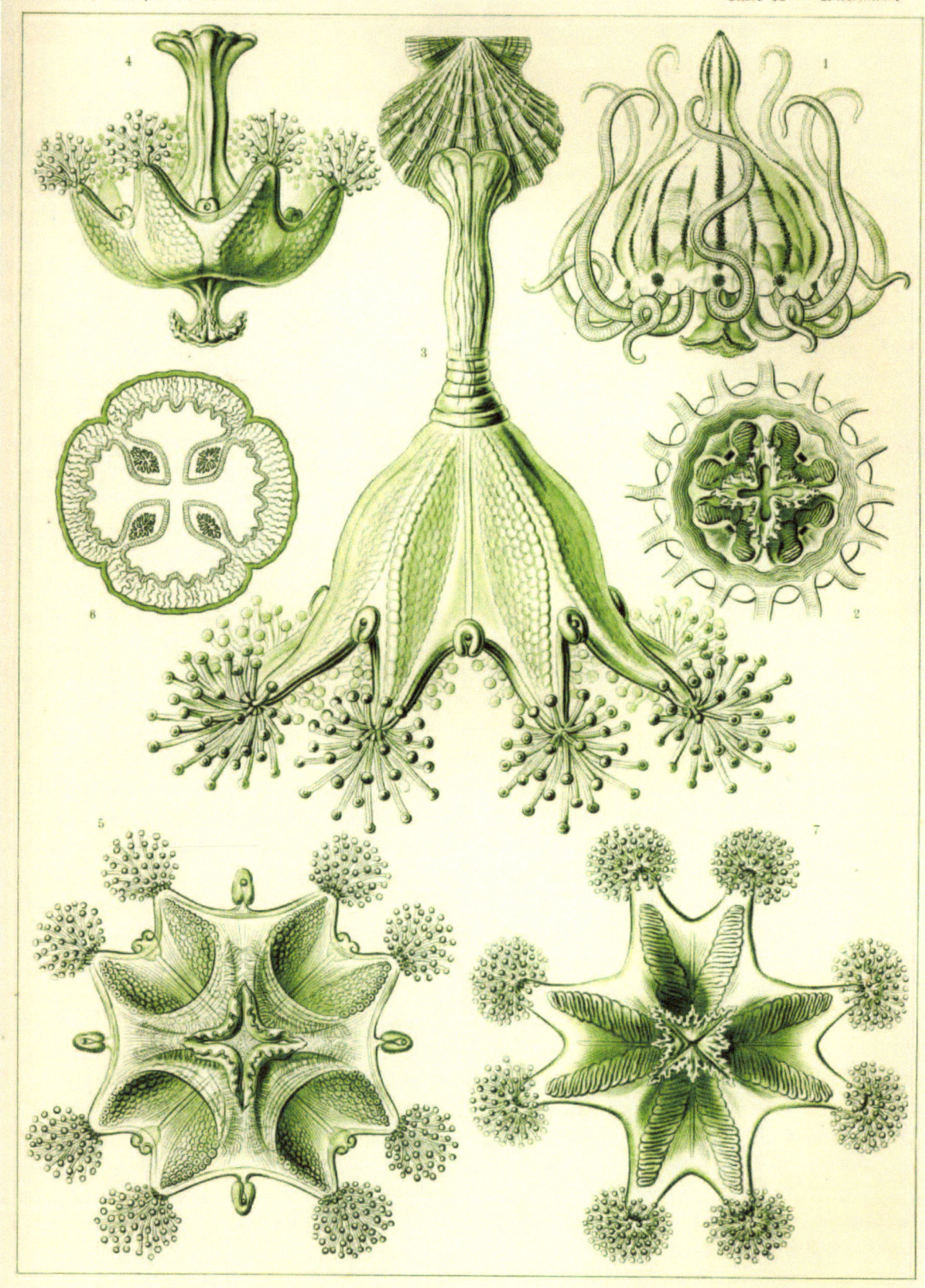

Stauromedusae. — Becherquallen.

Actiniae. — Seeanemonen.

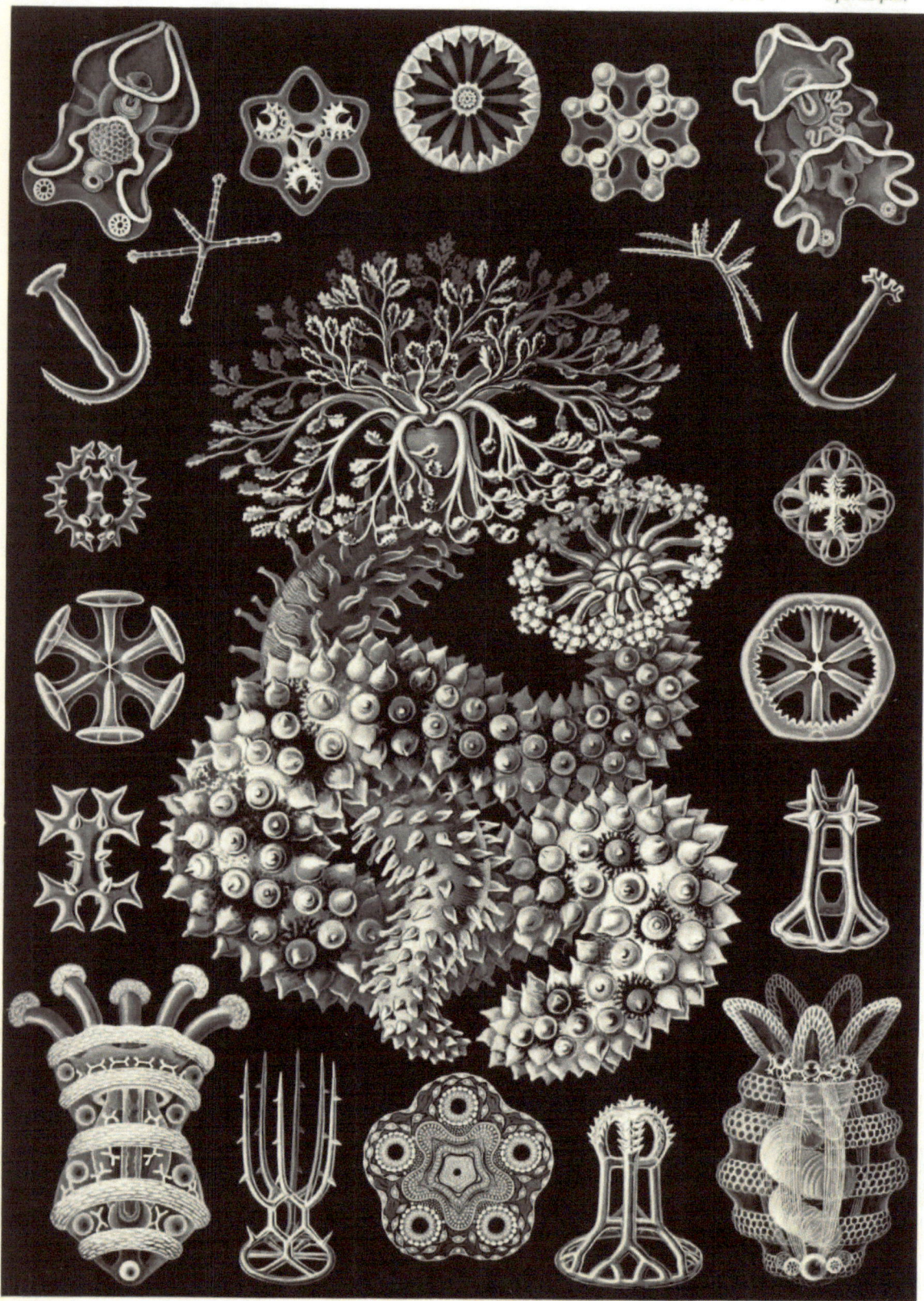

Thuroidea. — Gurkensterne.

Polycyttaria. — Vereins-Straßlinge.

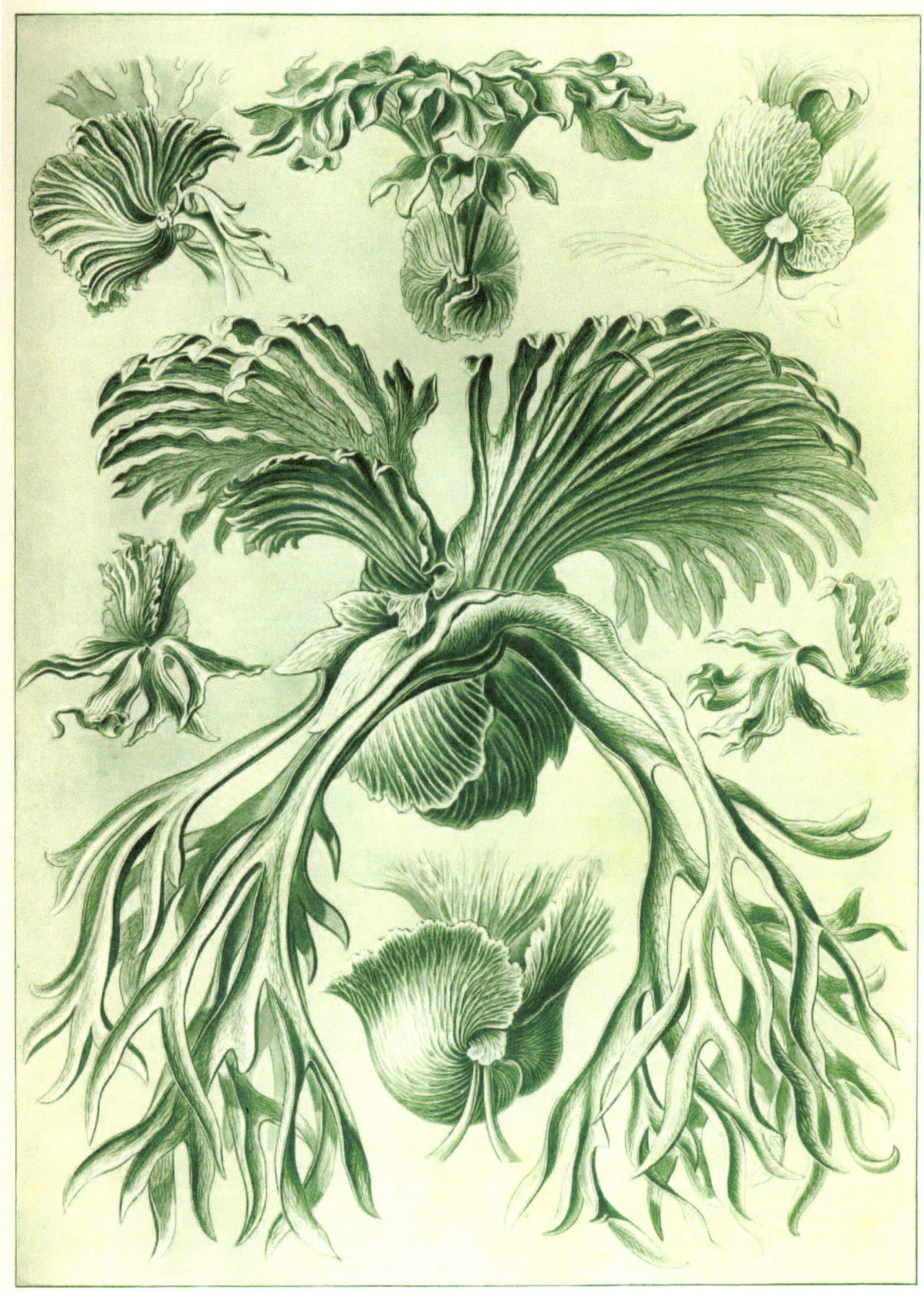

Filicinae. — Laubfarne.

Prosobranchia. — Vorderkiemen-Schnecken.

Gamochonia. — Trichterkraken.

Acephala. — Muscheln.

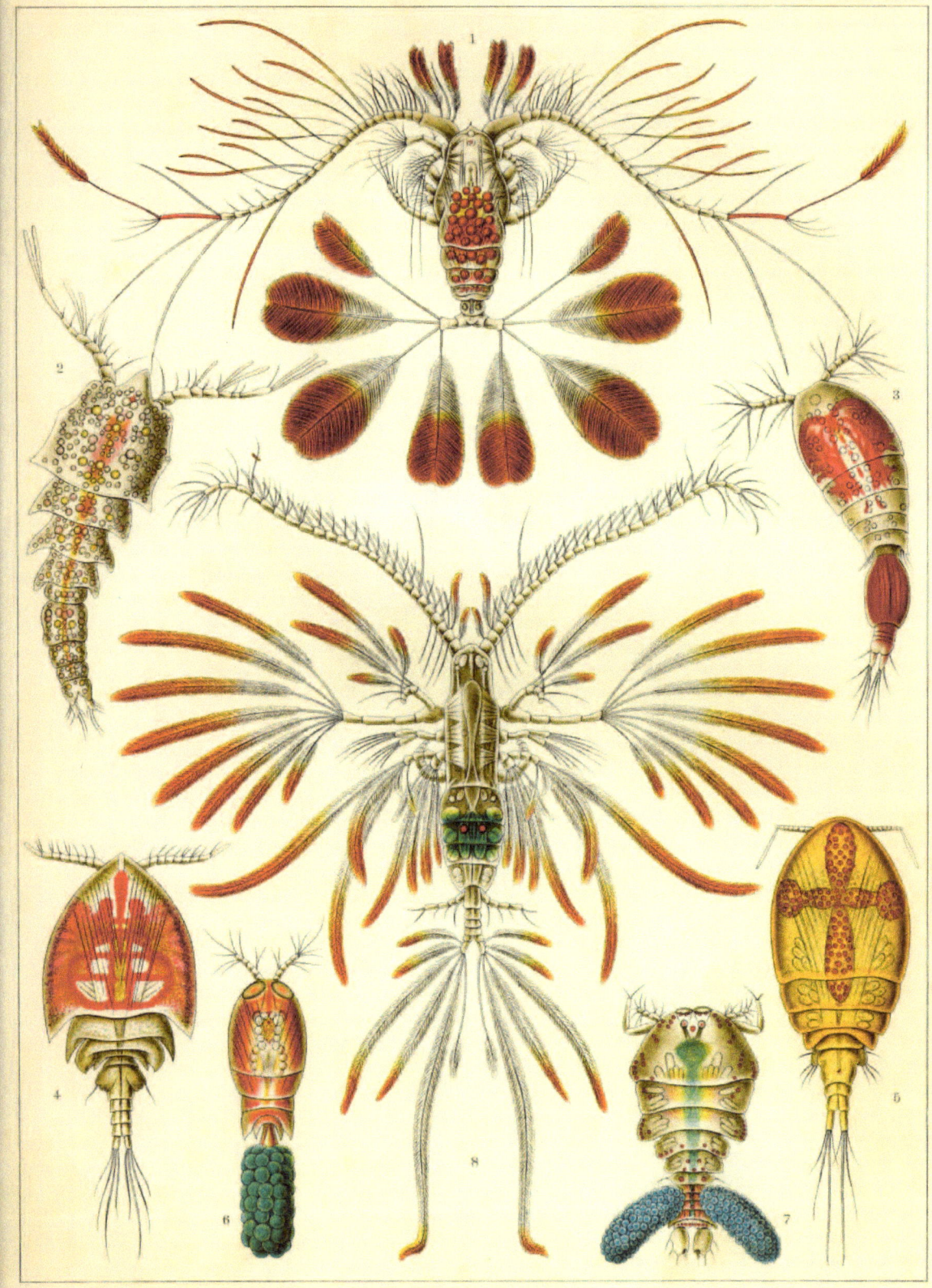

Copepoda. — Ruderkrebse.

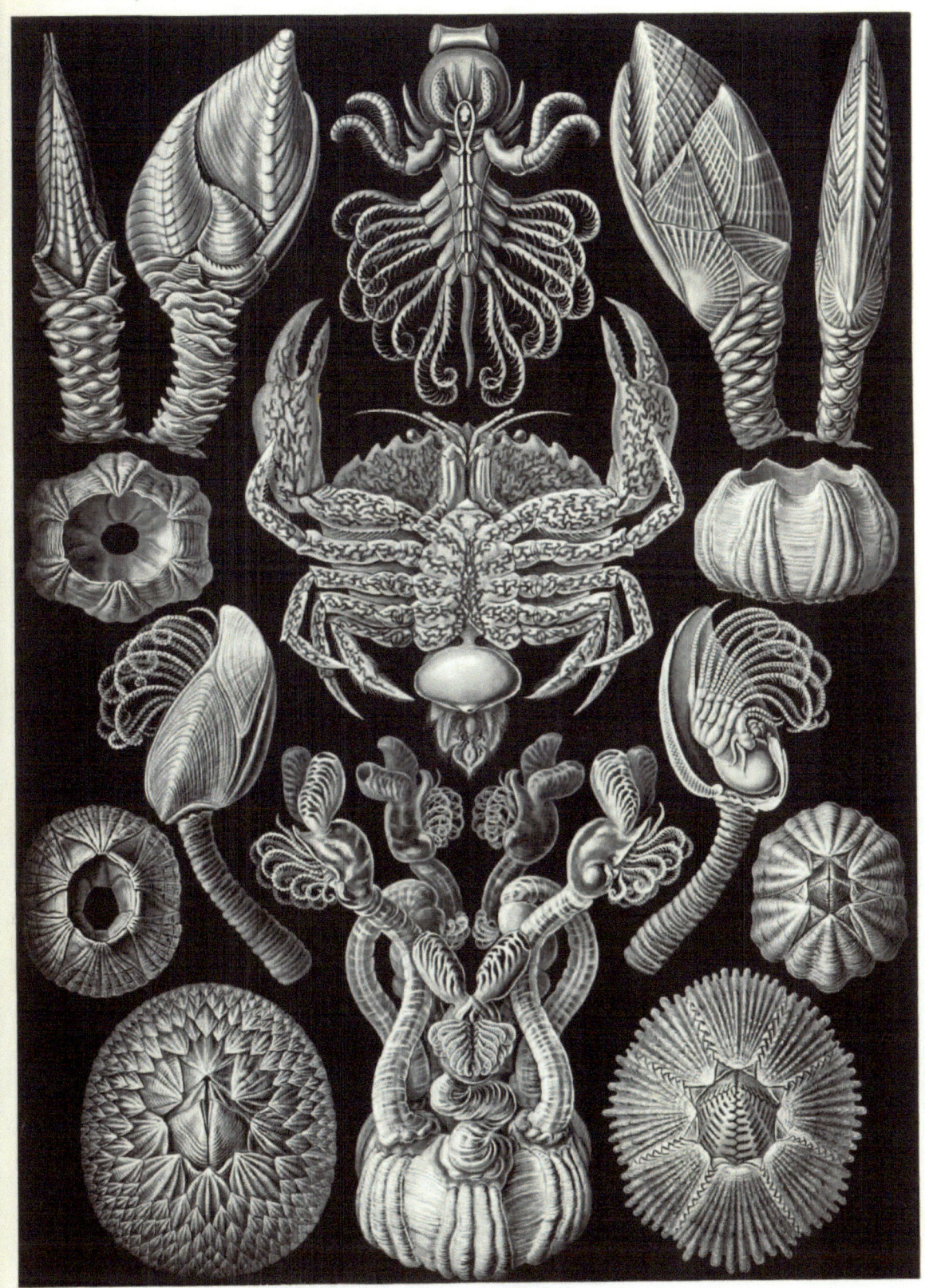

Cirripedia. — Rankenkrebfe.

Tineida. — Motten.

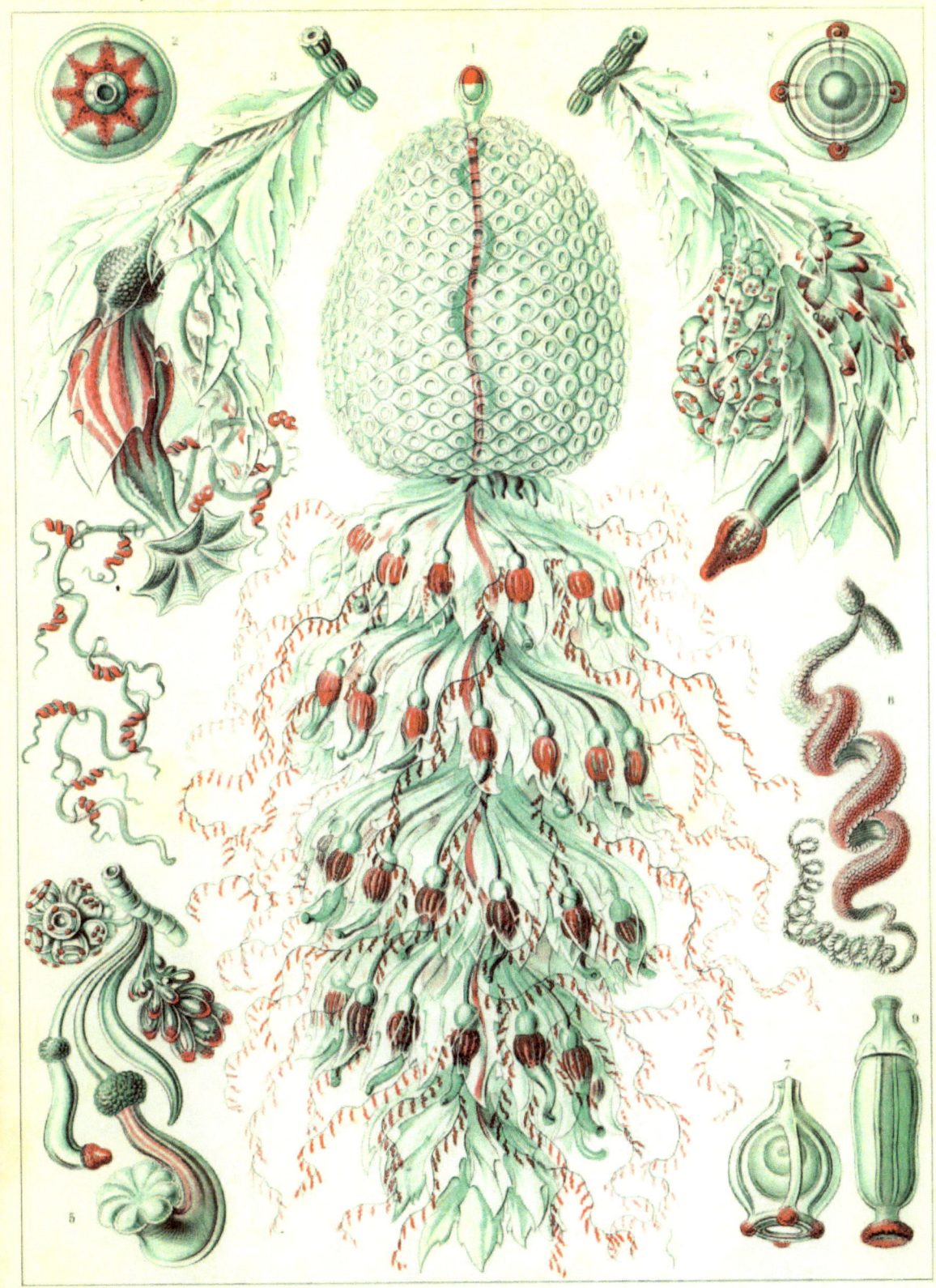

Siphonophorae. — Staatsquallen.

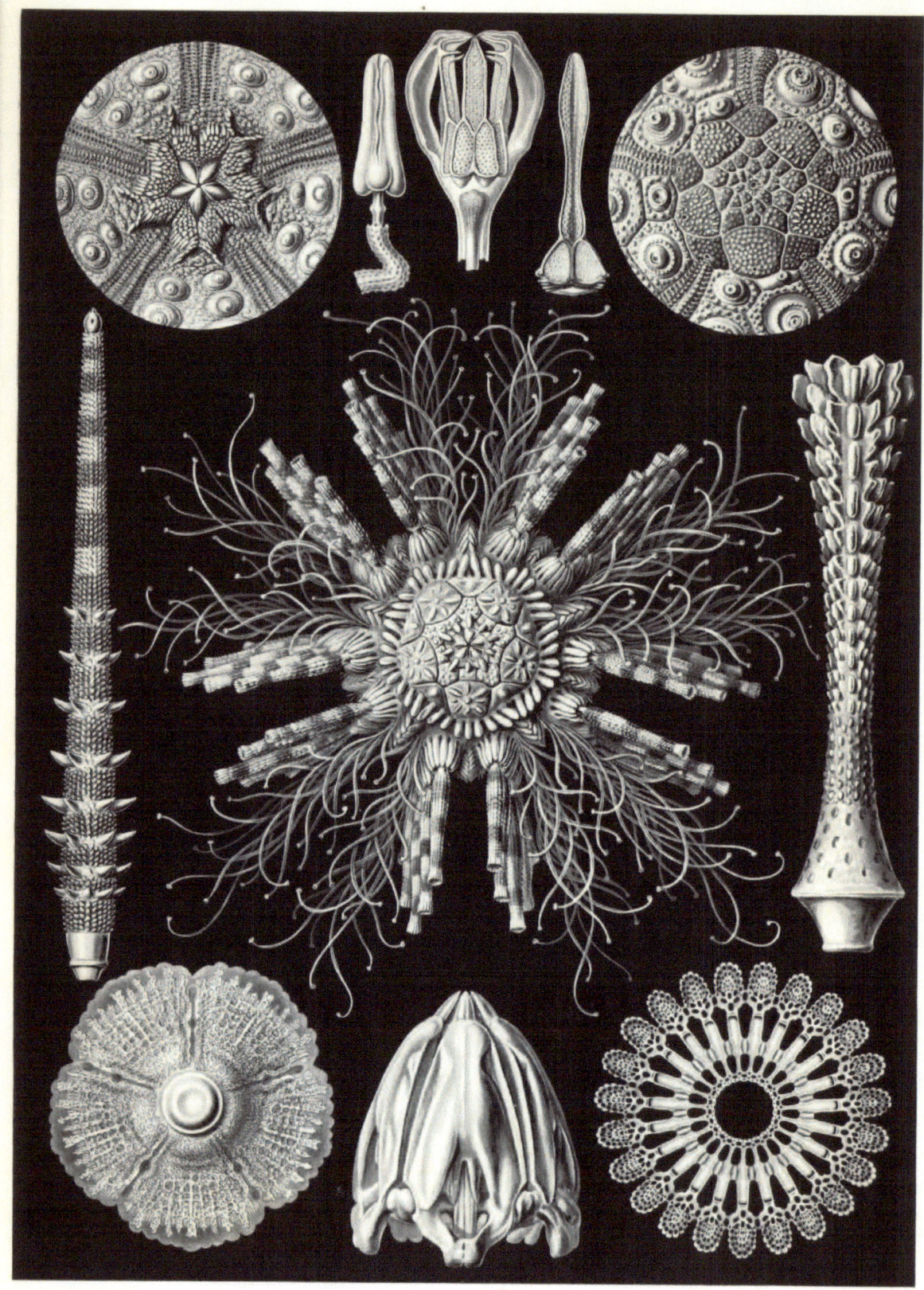

Echinidea. — Igelsterne.

Phaeodaria. — Rohrstrahlinge.

Nepenthaceae. — Kannenpflanzen.

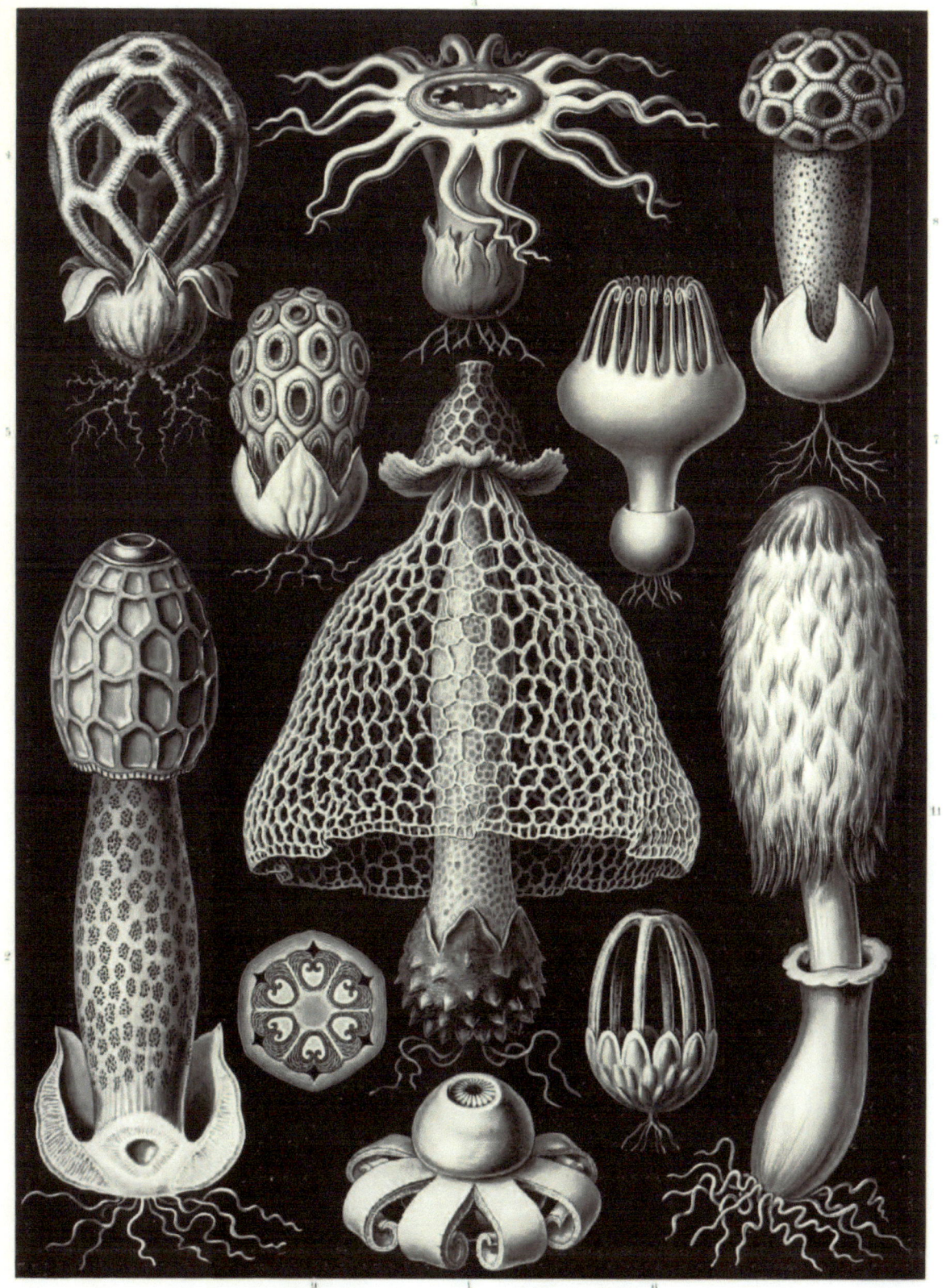

Basimycetes. — Schwammpilze.

Siphoneae. — Riesen-Algetten.

Florideae. — Rotalgen.

Arachnida. — Spinnentiere.

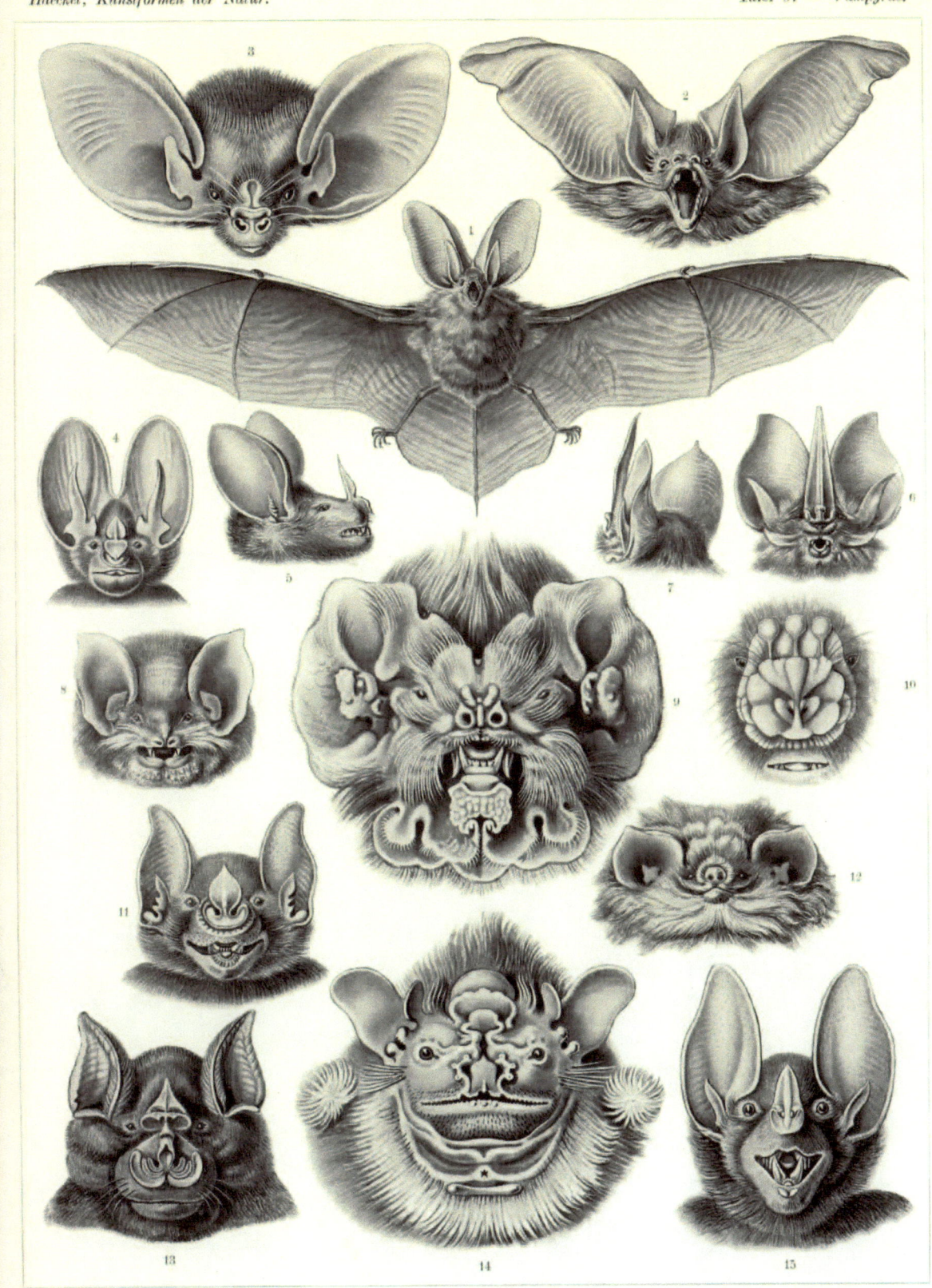

Chiroptera. — Fledertiere.

Batrachia. — Frösche.

Hexacoralla. — Sechsstrahlige Sternkorallen.

Ophiodea. — Schlangensterne.

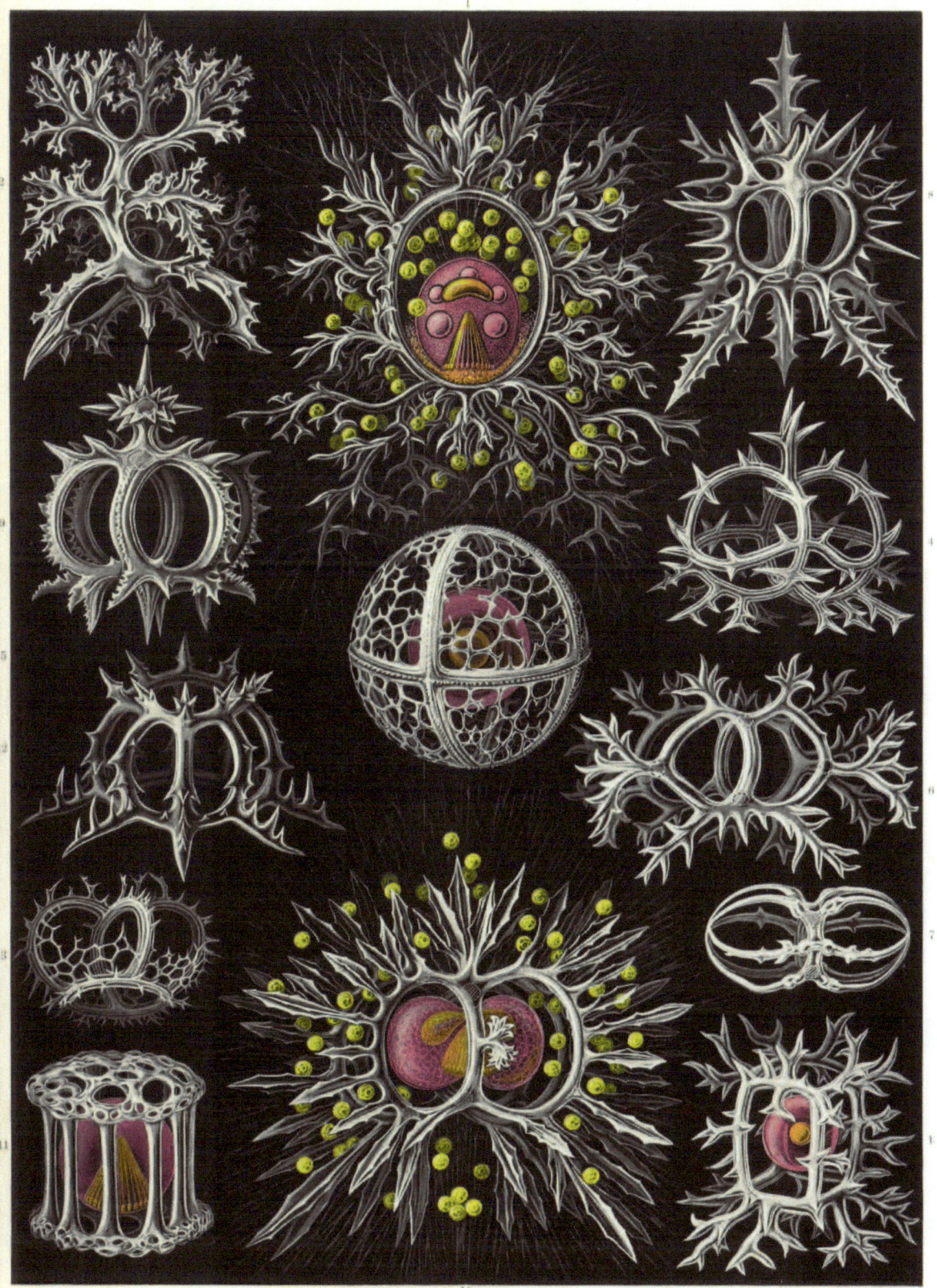

Stephoidea. — Ringelstrahlinge.

Muscinae. — Laubmoose.

Ascomycetes. — Schlauchpilze.

Orchideae. — Venusblumen.

Platodes. — Plattentiere.

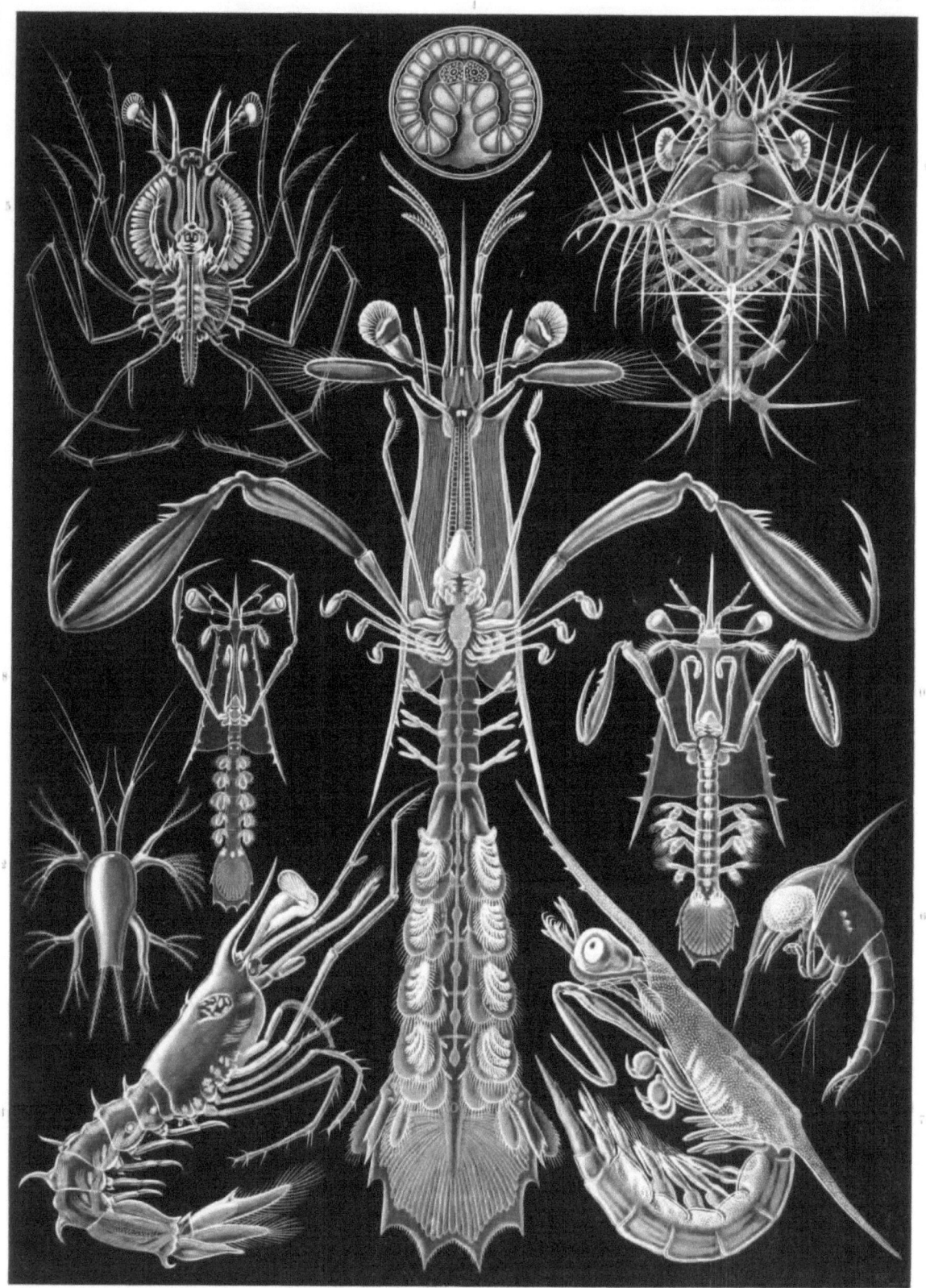

Thoracostraca. — Panzerkrebse.

Siphonophorae. — Staatsquallen.

Cubomedusae. — Würfelquallen.

Lacertilia. — Eidechsen.

Blastoïdea. — Knospensterne.

Thalamophora. — Kammerlinge.

Hepaticae. — Lebermoose.

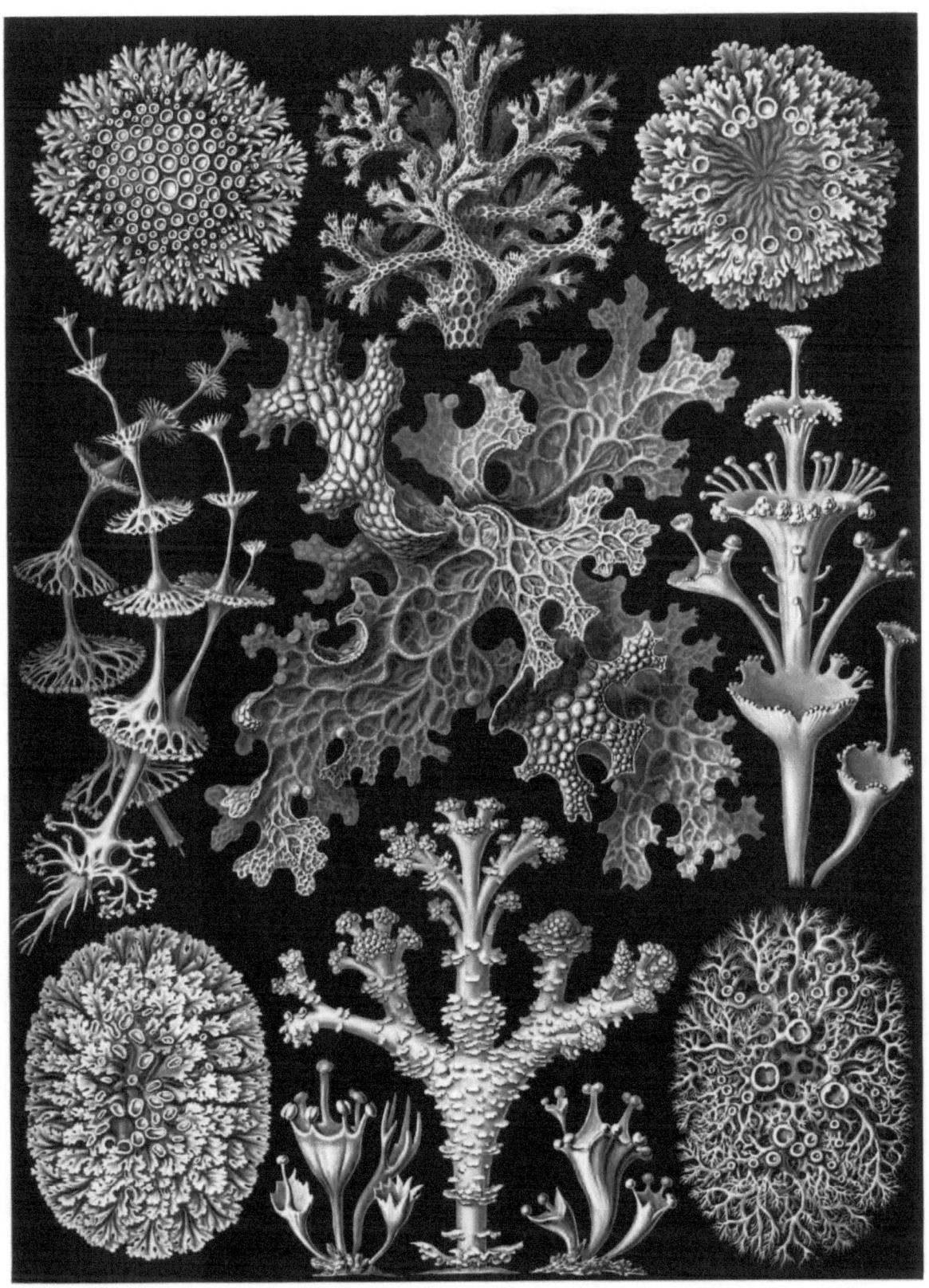

Lichenes. — Flechten.

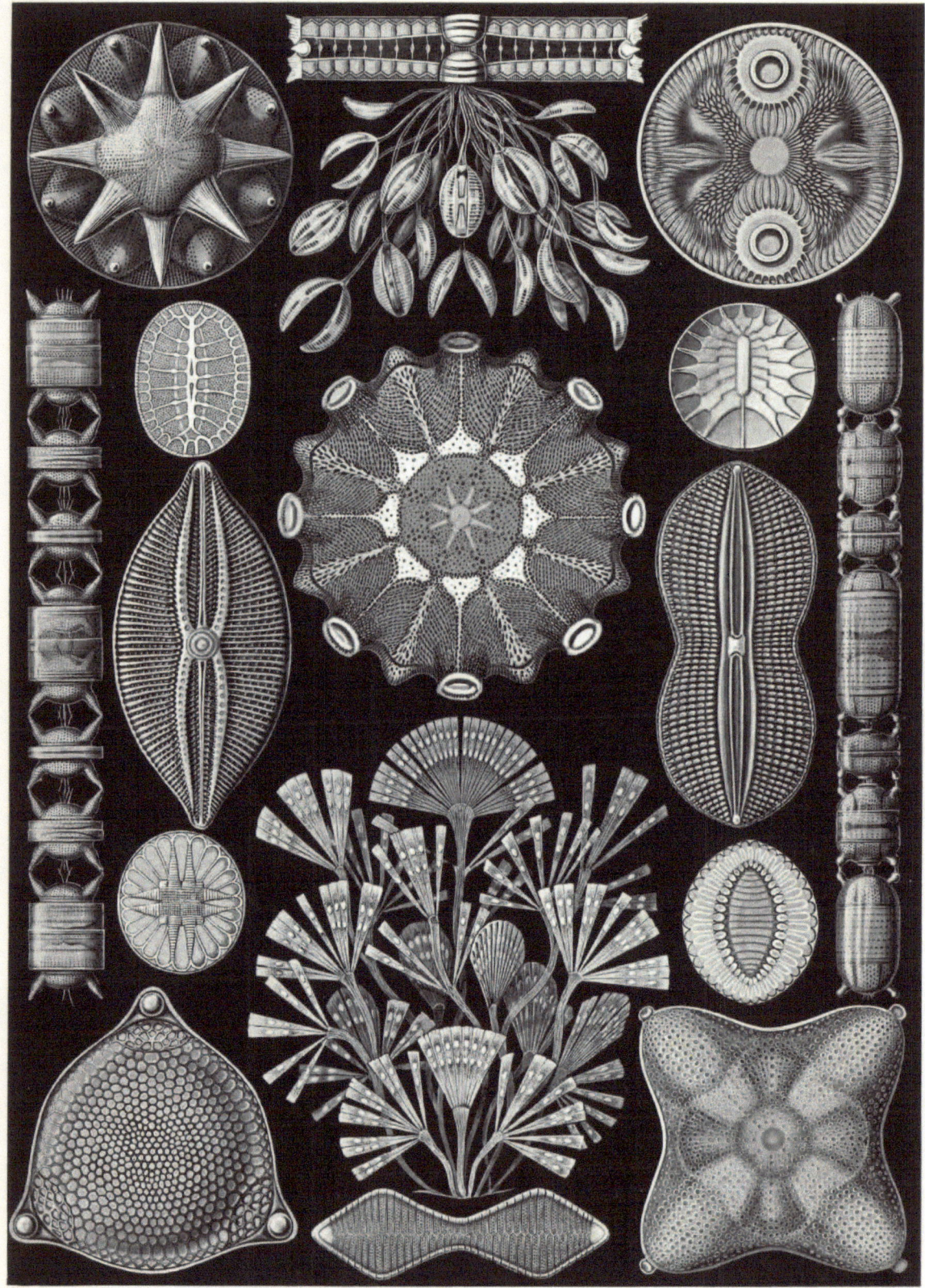

Diatomea. — Schachtellinge.

Ascidiae. — Seescheiden.

Decapoda. — Zehnfußkrebse.

Teleostei. — Knochenfische.

Discomedusae. — Scheibenquallen.

Chelonia. — Schildkröten.

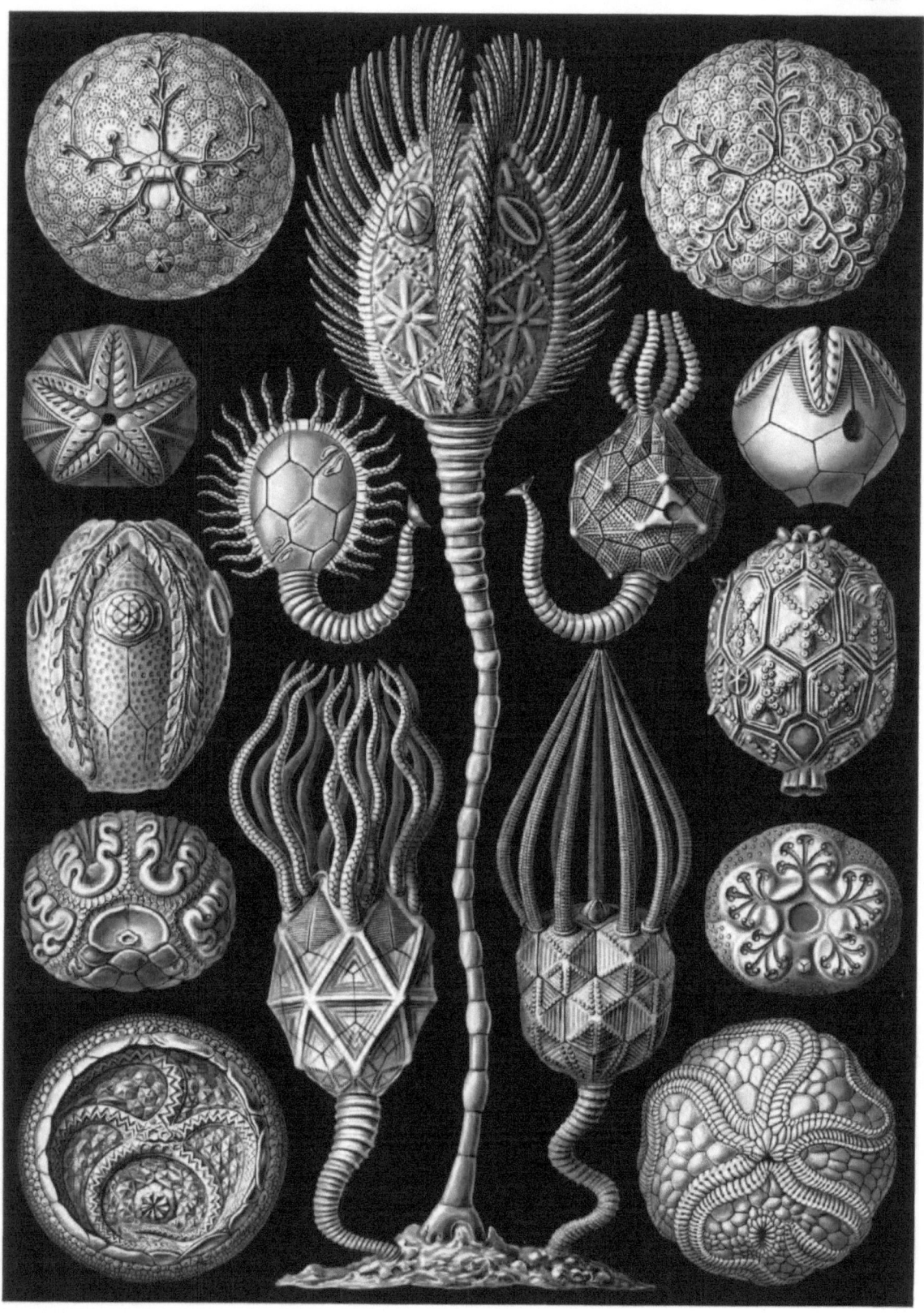

Cystoidea. — Beutelsterne.

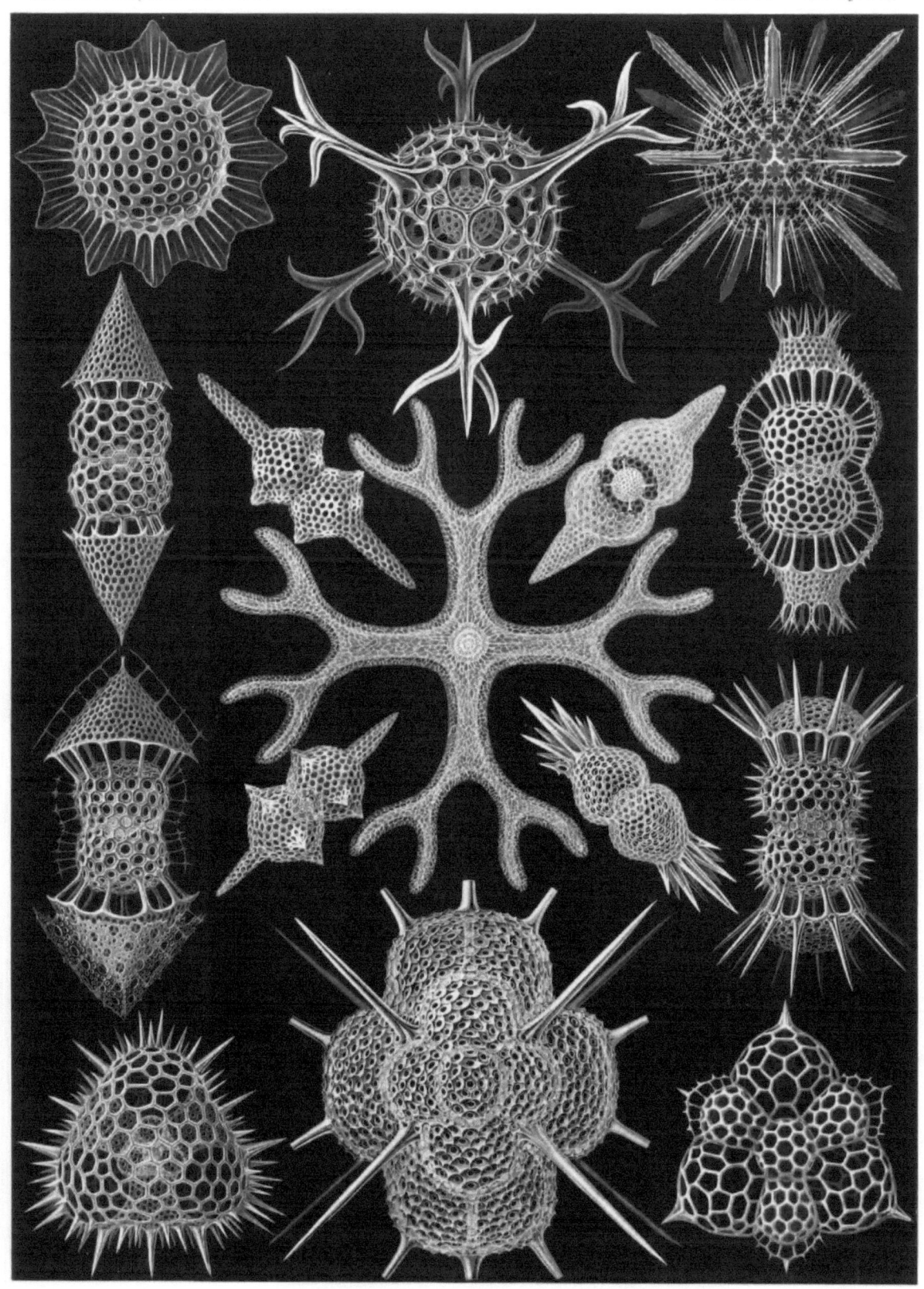

Spumellaria. — Schaumstrahlinge.

Filicinae. — Laubfarne.

Mycetozoa. — Pilztiere.

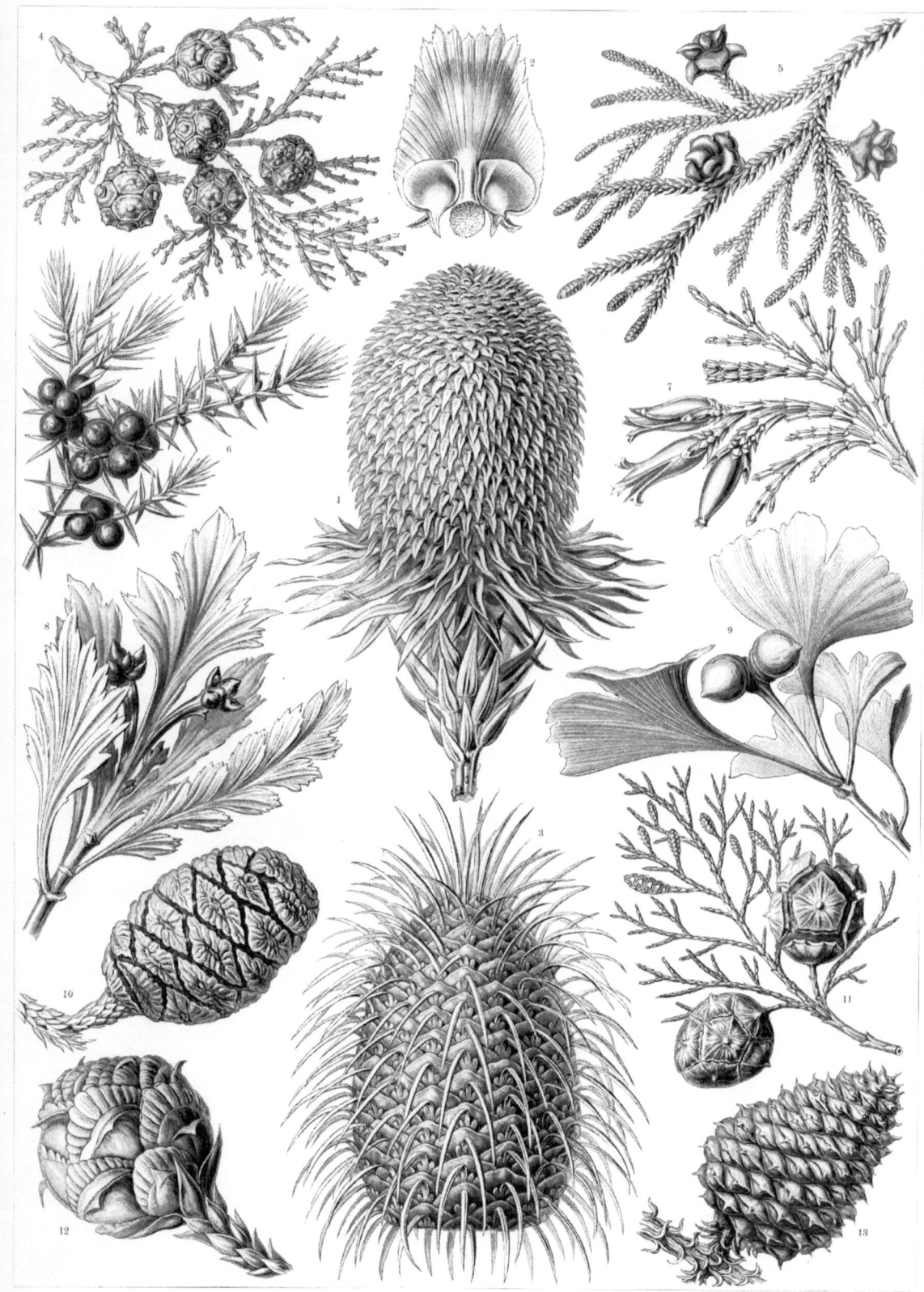

Coniferae. — Zapfenbäume.

Amphoridea. — Urnensterne.

Chaetopoda. — Borstenwürmer.

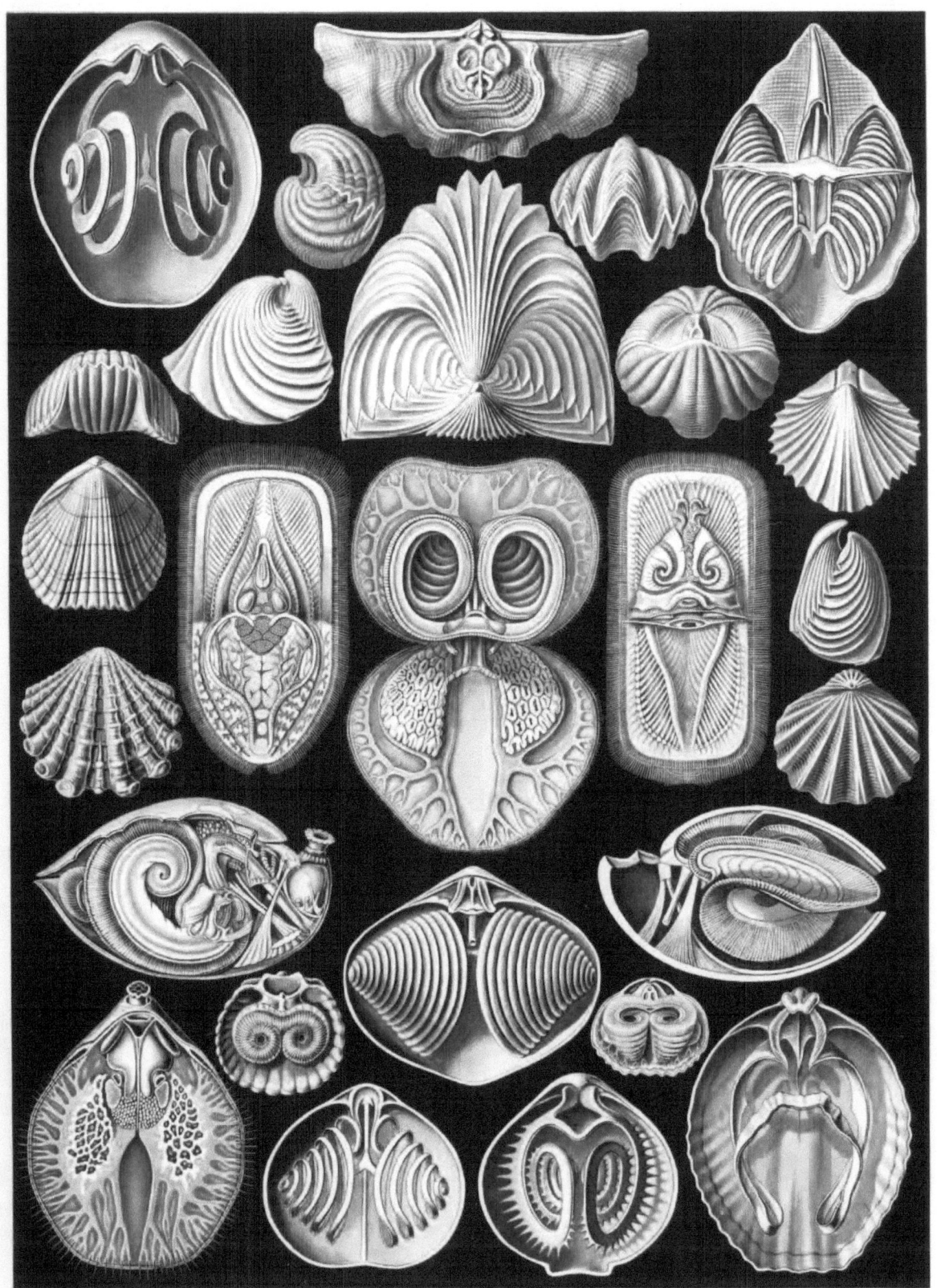

Spirobranchia. — Spiralkiemer.

Discomedusae. — Scheibenquallen.

Trochilidae. — Kolibris.

Antilopina. — Antilopen.

www.ingramcontent.com/pod-product-compliance
Lightning Source LLC
Chambersburg PA
CBHW050728180526
45159CB00003B/1162